Quantum Mechanics 1: Foundations

N. J. B. Green

Reader in Chemistry,
King's College, London

Series sponsor: **ZENECA**

ZENECA is a major international company active in four main areas of business: Pharmaceuticals, Agrochemicals and Seeds, Specialty Chemicals, and Biological Products.

ZENECA's skill and innovative ideas in organic chemistry and bioscience create products and services which improve the world's health, nutrition, environment, and quality of life.

ZENECA is committed to the support of education in chemistry and chemical engineering.

OXFORD   NEW YORK   TOKYO
OXFORD UNIVERSITY PRESS
1997

Oxford University Press, Great Clarendon Street, Oxford OX2 6DP

Oxford   New York
Athens   Auckland   Bangkok   Bogota   Bombay
Buenos Aires   Calcutta   Cape Town   Dar es Salaam
Delhi   Florence   Hong Kong   Istanbul   Karachi
Kuala Lumpur   Madras   Madrid   Melbourne
Mexico City   Nairobi   Paris   Singapore
Taipei   Tokyo   Toronto   Warsaw
and associated companies in
Berlin   Ibadan

Oxford is a trade mark of Oxford University Press

Published in the United States
by Oxford University Press Inc., New York

© N. J. B. Green, 1997

A catalogue record for this book is available from the British Library

Library of Congress Cataloging-in-Publication Data
(Data applied for)
ISBN   0 19 855761 2

Typeset by the author using LaTeX

Printed in Great Britain by Bath Press Ltd., Bath

Series Editor's Foreword

Oxford Chemistry Primers are designed to provide clear and concise introductions to a wide range of topics that may be encountered by chemistry students as they progress from the freshman stage through to graduation. The Physical Chemistry series aims to contain books easily recognized as relating to established fundamental core material that all chemists need to know, as well as books reflecting new directions and research trends in the subject, thereby anticipating (and perhaps encouraging) the evolution of modern undergraduate courses.

In this Physical Chemistry Primer, and its companion volume, Nick Green presents a clearly written and elegant account of the foundations and applications of Quantum Mechanics. He develops and builds on ideas that undergraduate chemists will have encountered via introductory courses to give insight and understanding of the behaviour of the microscopic world at the atomic and molecular level. This Primer will be of interest to all students of chemistry and their mentors.

Richard G. Compton
Physical and Theoretical Chemistry Laboratory
University of Oxford

Contents

0 Introduction

Quantum mechanics is a subject of central importance in chemistry. The chemist's approach to the understanding of matter and its chemical transformations is to take a microscopic view, connecting experimental observation with the properties of the constituent molecules. However, at the microscopic scale atoms and molecules and their constituent particles do not obey the familiar laws of mechanics that pertain in the everyday macroscopic world. For example, the existence of molecular line spectra indicates that molecules can be found only at certain well-defined energies, unlike objects at our own scale, such as tennis balls, whose energy appears to be continuously variable.

The aims of this primer and its companion are threefold:

1. to explain the fundamentals of quantum mechanics from the point of view of chemistry;

2. to describe the areas of chemistry where quantum mechanics is most important;

3. to show how quantum mechanics is applied to chemical problems.

The coverage is divided between two primers: the first deals with the foundations of the subject, concentrating on exactly soluble problems and methods for reducing complicated problems to simple formulations. The second part is a toolkit for applying quantum mechanics to chemical problems, and therefore concentrates on approximate methods. Part II will be referred to as II wherever appropriate.

The presentation is aimed at students in the later years of a chemistry degree; it is intermediate in difficulty between Atkins' *Physical chemistry* and *Molecular quantum mechanics*. Sections and problems which are more demanding are specifically identified. Quantum mechanics is by nature a mathematical theory: it is not possible either to develop or to use it without a certain amount of mathematics. We shall attempt to keep more advanced maths to a minimum. However, the reader is assumed to be familiar with the contents of a typical introductory *Mathematics for chemists* course, and any further concepts will be explained as and where they are needed.

1   Fundamental principles

1.1   Why do we need quantum mechanics?

As stated in the introduction, at the scale of atoms and molecules nature does not behave in the same way as macroscopic objects. This conclusion was reached in the first decades of this century following a number of rather surprising experimental results. Towards the end of the nineteenth century it had been thought that all the fundamental laws of nature were well-understood. Two distinct types of motion were known: wave motion and particle motion. Waves can be distinguished by the phenomenon of *interference* and are characterized by a frequency and a wavelength. Particles possess properties such as mass, position and momentum, and obey Newton's laws of motion. The two were clearly distinct. There had been a long-standing argument about the nature of light between the followers of Huygens, who believed it to be a wave motion, and those of Newton who maintained that it was corpuscular, but in 1801 James Young had resolved this argument in favour of Huygens by demonstrating the interference of light in his famous two-slit experiment, proving that light was a wave motion.

This state of affairs was so satisfactory that the Nobel laureate Michelson, when opening a new laboratory at the University of Chicago in 1894, stated 'our future discoveries must be looked for in the sixth decimal place'. He could hardly have been more wrong! The next 20 years saw the development of both quantum mechanics and relativity, the two greatest scientific revolutions since Newton. In fact the final triumphs of classical physics, which led to this view, also gave the first hints of trouble: the famous experiment in which Franck and Hertz proved that light was an electromagnetic wave, as postulated by Maxwell, also led to the discovery of the photoelectric effect, whose detailed properties cannot be explained on the basis of classical physics.

To demonstrate that something is a wave it is necessary to observe a diffraction pattern, which arises from the interference of waves.

Light comes in packets $h\nu$

By the turn of the century there were three separate experimental observations which were apparently inexplicable.

The ultra-violet catastrophe

All bodies in thermal equilibrium lose heat in the form of electromagnetic radiation with a characteristic spectrum, which depends on the temperature. For example, a human being emits radiation mainly in the infra-red region of the spectrum (leading to the possibility of infra-red imaging); hotter bodies glow red-hot and then white-hot, emitting visible light. The spectrum shifts to higher frequency as the body gets

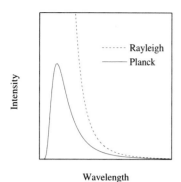

Fig. 1.1 Spectrum of thermal radiation at a temperature of 1500 K. The dashed line is the prediction of Rayleigh.

hotter, and empirical laws for the temperature dependence of the maximum of the spectrum (Wien's displacement law) and the rate of heat loss (Stefan's law) were discovered. In 1900 Lord Rayleigh calculated a theoretical spectrum from the laws of classical physics. The result bore little resemblance to the experimental spectrum: the calculated spectrum, depicted in Fig. 1.1, is proportional to T, has no maximum, increasing without limit at low wavelengths (in the ultra-violet), and predicts an infinite rate of heat loss. Rayleigh's calculation was correct according to the laws of physics accepted at the time. The problem is that these laws are not correct.

The photoelectric effect

When light shines on a metal surface electrons are ejected. This phenomenon was perfectly understandable in the context of classical physics; light causes electrons to resonate in the metal, and when enough energy has been absorbed by this resonance, electrons are lost. However, on closer examination three observed features of the process are incompatible with this simple idea.

1. No matter how low the intensity of the light, photoelectrons are ejected instantly, as soon as the light is switched on.

2. There is a critical cut-off frequency below which no electrons are emitted irrespective of the intensity.

3. If the maximum kinetic energy of the photoelectrons emitted is plotted against the frequency of the incident light, a straight line is obtained.

The first observation is problematical because in the classical picture, there should be a time delay during which the electrons build up enough energy to escape from the surface. The experiment seems to indicate that light energy is concentrated in small areas instead of being spread out uniformly over the wave front. The latter observations, summarized in Fig. 1.2, were puzzling because in Maxwell's theory the energy carried by the light should depend only on the amplitude of the wave and not at all on the frequency, and there should therefore be no connection between energy and frequency.

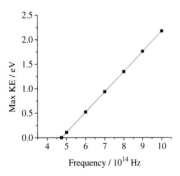

Fig. 1.2 Maximum kinetic energy of photoelectrons emitted from Cs metal as a function of light frequency.

Atomic line spectra

When atoms are excited (for example in a flame or an electric discharge) they emit light. Atoms of a particular element emit light in a line spectrum of sharply defined discrete frequencies. Classical physics offers no clue to the reason for this observation, apart from an analogy with the characteristic frequencies of sound emitted by a vibrating object such as a guitar string or an organ pipe. Since the characteristic frequencies of an atom do not in general obey a simple formula (apart from the special case of the H atom), whereas those of a guitar string do, the analogy is at best incomplete.

Planck's quantum theory

The resolution of the ultra-violet catastrophe was proposed in 1900 by Planck, who succeeded in explaining the form of the black-body spectrum, but at the expense of abandoning the classical picture of light as an electromagnetic wave. Planck showed that if light energy is concentrated in packets of energy $E = h\nu$ (ν is the frequency of the light), and if the Boltzmann distribution can be applied to the number of such packets (or *quanta*) at each of the characteristic frequencies of the spectrum calculated by Rayleigh, then the ultra-violet catastrophe is removed. The reason for this is that at high frequencies a quantum contains so much energy that it is highly improbable for there to be even one quantum in a given frequency interval, according to the Boltzmann distribution. More detailed analysis shows not only that the ultra-violet catastrophe is avoided, but also that the calculated spectrum is in excellent agreement with experiment.

Although Planck's theory of the black-body spectrum is highly successful, it comes at the expense of introducing a new fundamental physical constant h and of overthrowing the classical picture of light. It was obviously necessary to have corroborative evidence before such a radical new theory could be accepted. This corroboration was provided by Einstein's use of the same idea to explain the problems of the photoelectric effect. The localization of light energy in packets of size $h\nu$ (*photons*) leads to a ready explanation of the immediate ejection of photoelectrons. The threshold frequency below which no photoemission is observed can also be explained easily, since below this frequency photons do not have enough energy to overcome the binding energy of the electrons in the metal (the *work-function*), ϕ. Furthermore, the law of the conservation of energy states that the maximum kinetic energy of the photoelectrons is equal to $h\nu - \phi$, so that a plot of this maximum kinetic energy against the frequency of the incident light is a straight line of slope h, as shown in Fig. 1.2. The final clincher is that the value of h obtained from such a plot is the same as that needed to fit the observed black-body spectrum, implying that h is indeed a new fundamental constant. When combined with Planck's hypothesis, the discrete frequencies found in atomic line spectra imply that the atom can only lose or gain energy in certain discrete amounts. It is therefore likely that atoms can only exist with certain well-defined internal energies; the spectrum comes about when the atom falls from one such state to another, emitting the energy difference as a photon, $E_2 - E_1 = h\nu$. This formula is known as Bohr's frequency condition. The idea of discrete energy levels gained further credence when Bohr subsequently calculated the allowed energy levels of the hydrogen atom and explained its spectrum quantitatively.

Photons carry momentum $p = h/\lambda$

The idea that light sometimes behaves as if it were made up of particles leads immediately to the question of whether the photon has any other properties characteristic of particles. Since energy is related to mass by

To show that a photon has a momentum, Compton showed that it scatters off matter obeying the principles of the conservation of momentum and energy.

In the *laser cooling* experiment, atoms in a velocity-selected beam absorb photons from a counter-propagating laser beam; the atoms slow down, and eventually become approximately stationary.

To show that electrons have wavelike behaviour it is necessary to demonstrate a diffraction pattern arising from their interference.

the equation $E = mc^2$, the photon (travelling at the speed of light, c) has a relativistic mass $m = h\nu/c^2$. This has been verified experimentally by observing how light from distant stars is deviated by the gravitational field of the sun. However, the photon is rather a special case, since it only has any existence when travelling at the speed of light. Since a photon has both mass and velocity, it should also have a momentum $p = mc = h\nu/c = h/\lambda$, which is inversely proportional to the wavelength of the light, λ. This relation was first proposed by de Broglie (1923).

The de Broglie relation was verified by the discovery of the *Compton effect*—the relationship between the angle at which X-rays scatter off electrons in matter and the shift in the wavelength of the scattered X-ray. This relationship can be derived on the assumption that the photon behaves as a particle, scattering off an electron in the same way as a moving billiard ball scatters off a stationary billiard ball (i.e. from the principles of the conservation of energy and momentum[1]). More recently the momentum of the photon has been demonstrated by measurement of the *radiation pressure* inside a laser, which arises from photons bouncing back and forth between the two mirrors in the laser producing a force on the mirrors. Another graphic experiment relying on the momentum of the photon is the deflection of a particle beam by successively absorbing photons from a laser and then re-emitting them in random directions. It has been suggested that this technique might be a viable means of separating uranium isotopes, since isotopically substituted molecules absorb light of slightly different frequencies, so that in a beam of molecules, only those which absorb the incident light will be deflected.

Moving particles have a characteristic wavelength

de Broglie not only suggested that photons carry a characteristic momentum h/λ, but also that the same relationship might hold for all moving particles, i.e. that a particle with momentum p should have a characteristic wavelength $\lambda = h/p$. The truth of this extraordinary suggestion was proved by Davisson and Germer and by G. P. Thompson who diffracted a beam of electrons from a metal foil and showed that the interference pattern obtained was characteristic of the de Broglie wavelength $\lambda = h/p$.

Subsequent experiments showed that He atoms and H_2 molecules also diffract to give characteristic interference patterns. Nowadays both electron diffraction and neutron diffraction are widely used in investigating the structures of molecules, surfaces and condensed matter.

1.2 Representation of a particle by a wavefunction

If a particle with a fixed momentum has a fixed characteristic wavelength then it must be possible to represent it mathematically in the same way as a wave is represented, i.e. by using a *wavefunction*. In classical physics a steady wave motion of wavelength λ propagating in the positive x-direction is represented by the equation

$$\psi(x) = \exp(\mathrm{i}2\pi x/\lambda) = \cos(2\pi x/\lambda) + \mathrm{i}\sin(2\pi x/\lambda) \qquad (1.1)$$

where i denotes the square root of -1. Substituting the de Broglie relation $\lambda = h/p$ this becomes

$$\psi(x) = \exp(\mathrm{i}px/\hbar) \qquad (1.2)$$

where $\hbar = h/2\pi$. This complex form of the wave distinguishes between a wave propagating in the positive direction (i.e. a particle with positive momentum p) and one propagating in the negative x-direction (i.e. with momentum $-p$), which can be obtained simply by changing p to $-p$ in eqn 1.2.

In three dimensions the wavefunction for a particle with momentum vector **p** is $\exp(\mathrm{i}\mathbf{p}.\mathbf{r}/\hbar)$, where **p.r** represents the scalar product of **p** with position vector **r**.

Wavefunctions and probability

It is all very well to be able to represent a moving particle with a wavefunction, but it is frequently necessary to calculate other quantities from the wavefunction, for example the location of the particle or its energy. The relationship between the wavefunction and the particle location is probabilistic in nature. In a light beam the wavefunction is spread out evenly over the whole wave front, but yet when the photons are detected, for example on a fluorescent screen, they arrive at unpredictable locations with a probability density proportional to the intensity of the beam or to the square of the amplitude of the wave. Born suggested that this should also be the case for wavefunctions representing other particles. The probability density for the location of a particle represented by a wavefunction ψ should be the square of the amplitude of the wavefunction, $|\psi|^2$. To make this idea precise in one dimension we say that if a particle is represented by the wavefunction $\psi(x)$ the probability of its being found in the interval between x and $x + \mathrm{d}x$ is

$$|\psi(x)|^2 \mathrm{d}x = \psi^* \psi \, \mathrm{d}x \qquad (1.3)$$

where ψ^* represents the complex conjugate of ψ.

In three dimensions the probability density of the particle is $|\psi(\mathbf{r})|^2$, which means that the probability of finding the particle in the infinitesimal element of volume $\mathrm{d}\tau = \mathrm{d}x\,\mathrm{d}y\,\mathrm{d}z$ at the point **r** is $|\psi(\mathbf{r})|^2\mathrm{d}\tau$. In an exactly analogous way, if $\rho(\mathbf{r})$ is the mass density of a nonuniform body then the mass contained in the volume element $\mathrm{d}\tau$ at the point **r** is $\rho(\mathbf{r})\,\mathrm{d}\tau$. The analogy between mass density and probability density can be continued. If it is desired to calculate the mass of a part of the body, then we sum all the infinitesimal mass elements $\rho(\mathbf{r})\,\mathrm{d}\tau$ that lie in that part of the body, by integration of the density. In the same way if it is desired to calculate the probability of finding the particle in a certain region of space the probability density must be integrated over that region. In particular integrating the mass density over the whole body gives the mass of the whole body, and integrating the probability density over the whole of the allowed space gives the probability of finding the particle somewhere in the allowed space, which must be unity.

This property of a probability density function is called *normalization*. For a normalized wavefunction

$$\int |\psi|^2 \, d\tau = 1 \qquad (1.4)$$

Worked example. Consider a particle confined to move on the x-axis between $x = 0$ and $x = a$ with a wavefunction $\psi(x) = N\sin(\pi x/a)$.

(i) Normalize the wavefunction, and

(ii) calculate the probability of finding the particle in the central half of the box, between the points $x = a/4$ and $x = 3a/4$.

Solution. (i) We are required first to *normalize* the wavefunction, which means to find a value of the constant N to ensure that the probability of finding the particle somewhere between $x = 0$ and $x = a$ is unity. To do this we consider the integral of the probability density over the whole of the allowed space, which must equal unity.

$$1 = \int_0^a N^2 \sin^2(\pi x/a) dx = N^2 a/2 \qquad (1.5)$$

Hence $N = \sqrt{2/a}$. [The integral can be done using the trigonometric identity $\sin^2 t = \frac{1}{2}(1 - \cos 2t)$ or else by inspection, noting that the function $\sin^2 t$ is symmetrically disposed about the value $\frac{1}{2}$.] (ii) Next we are required to calculate the probability of finding the particle between $a/4$ and $3a/4$. This is given by the integral of the probability density over the required interval, sketched in Fig. 1.3:

$$P = \frac{2}{a} \int_{a/4}^{3a/4} \sin^2(\pi x/a) dx = \frac{1}{2} + \frac{1}{\pi} \qquad (1.6)$$

[The integral can be performed using the same trigonometric identity.]

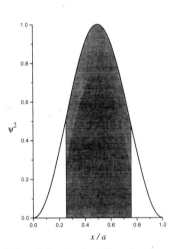

Fig. 1.3 Probability density function. The integral covers the shaded region.

The superposition principle

The major distinguishing feature of wave motion is the phenomenon of interference. It was the observation of an interference pattern which enabled Young to prove that light was a wave motion, and it was the observation of an interference pattern from electron diffraction (the Davisson–Germer experiment) which verified the de Broglie relation for particles in motion as well as for light, leading to the introduction of the wavefunction. Interference arises from the superposition of two or more simple waves propagating through the same space. The principle of superposition is one of the most important principles of quantum mechanics. It can be understood in the following way. Suppose that if the wavefunction of the system is ψ_1 then some measurement will yield a particular result R_1 with probability one; similarly if the wavefunction is ψ_2 the measurement will certainly give a second result R_2. Then a similar measurement of any system whose wavefunction has the form $c_1\psi_1 + c_2\psi_2$, in which

c_1 and c_2 are constants, will yield either result R_1 or result R_2 (and no others). The linear combination $\psi = c_1\psi_1 + c_2\psi_2$ is considered to be a *superposition* or interference of the two functions ψ_1 and ψ_2. The superposition function is obtained simply by adding the two waves $c_1\psi_1$ and $c_2\psi_2$, and the resulting interference pattern is the spatial probability distribution, which is obtained in the usual way as $\psi^*\psi$. Superposition only has an observable effect if the two wavefunctions occupy the same region of space, otherwise there is no interference between them.

Young's two slit experiment

As an example of a superposition of two wavefunctions we consider the two-slit diffraction experiment, shown schematically in Fig. 1.4. Peaks of the waves propagating from each slit (marked x) are denoted by full lines, and troughs by dotted lines. This is the experiment by which Young originally demonstrated that light was a wave motion. Thus ψ_1 is the wavefunction for the light which passes through slit 1, and ψ_2 is that for light passing through slit 2. If the wavefunction is given by ψ_1 then a measurement (R_1) would indicate that the light passes through slit 1 with certainty (likewise for ψ_2 and slit 2). If both slits are open then the light passes with equal amplitude through each slit and the resulting wavefunction is a combination or superposition of the two wavefunctions $\psi = \psi_1 + \psi_2$. The resulting pattern ($R_{12} = \psi^*\psi$), projected onto a screen consists of a series of peaks of high intensity separated by regions of very low intensity (Figure 1.5). Peaks are found in directions where waves coming through the two slits reinforce one another, i.e. directions where peaks in the function ψ_1 coincide with peaks in the function ψ_2. These directions are indicated by arrows in Fig. 1.4. Likewise low intensity is observed in directions where peaks in ψ_1 coincide with the troughs in the wave ψ_2 so that the two waves interfere destructively cancelling each other out.

It is instructive to separate the interference pattern in two parts:

$$\psi^*\psi = (\psi_1^* + \psi_2^*)(\psi_1 + \psi_2) = [\psi_1^*\psi_1 + \psi_2^*\psi_2] + [\psi_1^*\psi_2 + \psi_2^*\psi_1] \quad (1.7)$$

The first term $[\psi_1^*\psi_1 + \psi_2^*\psi_2]$ represents the sum of the probability distribution that would be obtained simply by adding the distribution found when the second slit is blocked $[R_1 = \psi_1^*\psi_1]$ and that found when the first slit is blocked $[R_2 = \psi_2^*\psi_2]$. The second term is the interference between ψ_1 and ψ_2; in some directions this term is positive, leading to a reinforcement of intensity, and in some directions it is negative, leading to a diminution of intensity. In the case under consideration explicit formulas can be found for each term relatively easily.[2]

Mathematical properties of the wavefunction

Not every mathematical function is suitable as a wavefunction, the function must be 'well-behaved' in the sense of obeying a number of mathematical constraints.

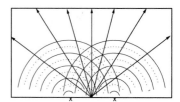

Fig. 1.4 Interference of waves in the two slit experiment. The two slit positions are marked x. The full lines represent peaks of the wave motion and the dashed lines represent troughs. The arrows indicate approximate directions of constructive interference.

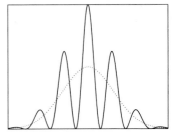

Fig. 1.5 A typical two-slit interference pattern. The dashed line is the classical result without interference (the first term of eqn 1.7). The full line is the observed pattern including interference.

1. ψ must be finite (since $|\psi|^2$ represents a probability density).

2. ψ must be single-valued.

3. ψ must be continuous.

4. ψ must be differentiable twice.

5. $|\psi|^2$ must be integrable over all space (otherwise ψ cannot be normalized). This implies that ψ and its derivatives must tend to zero sufficiently fast as $|\mathbf{r}| \to \infty$ in all directions.

In addition we can define two other important terms here: a wavefunction is said to be *normalized* if

$$\int \psi^* \psi \, d\tau = 1 \tag{1.8}$$

and two nonzero wavefunctions ψ_1 and ψ_2 are *orthogonal* if

$$\int \psi_1^* \psi_2 \, d\tau = 0 \tag{1.9}$$

which can be interpreted to mean that there is exactly zero overlap between the two wavefunctions.

1.3 Operators, observables, eigenfunctions and eigenvalues

The only information we can obtain about a microscopic physical system is contained in its wavefunction. We have just seen how the wavefunction is manipulated to give the spatial probability distribution of the system it represents. However the location of the particle is not the only quantity of interest, and we also need to be able to find out what information ψ contains about other mechanical quantities, such as momentum, energy or angular momentum. The wavefunction itself is not directly measurable, but all of these mechanical quantities are in principle measurable, and are termed *observables*. There are several formulations of quantum mechanics, all of which are mathematically equivalent.[3,4] Here we introduce only the most commonly used version, in which observables are represented mathematically by *operators*.

An operator is a mathematical object which acts on a function, transforming it into another function. Although the concept of an operator may not be familiar, the reader will undoubtedly have used operators in some form before. For example, the operator $\hat{\partial}_x$, meaning 'differentiate with respect to the variable x', operates on a function $f(x)$ transforming it into its derivative. In this book operators are distinguished from other mathematical objects by wearing a hat ($\hat{}$). The effect of an operator is to transform a function into a new function; for example, the derivative of a function f is not in general the same function as f, (e.g. $\hat{\partial}_x x^2 = 2x$). However, for a given operator it may be possible to find a family of functions which remain unchanged under its action (apart from possibly being multiplied by a constant), in other words functions which are transformed into themselves by the operator. For example, the operator $\hat{\partial}_x$ transforms the function $\exp(\alpha x)$ into $\alpha \exp(\alpha x)$, which is the same function multiplied by a constant factor α. A function which is transformed into itself by an operator is called an *eigenfunction* of the operator, and the constant multiplication factor introduced by the operation is called an *eigenvalue*. The function $\exp(\alpha x)$ is an eigenfunction of the operator $\hat{\partial}_x$, and the corresponding eigenvalue is α.

The idea of an eigenfunction and its eigenvalue is of great importance in quantum mechanics since all observables are represented by operators, and the eigenvalues of an operator are the only values that the corresponding observable may take. Any measurement of the observable must find one of the possible eigenvalues. For example, the permitted electronic energy levels of an atom are eigenvalues of the energy operator; any measurement to determine the electronic energy of an atom must find the atom at one of these energy levels.

The correspondence principle

In quantum mechanics all physical observables have corresponding operators. The formulation is based on the identification of two fundamental operators, from which all others may be constructed. These two special operators represent the position and the momentum of a particle.

If the motion of the particle is limited to one dimension (for example along the x-axis) the two operators are

$$x - \text{coordinate: } \hat{x} = x$$
$$x - \text{momentum: } \hat{p}_x = -i\hbar\frac{\partial}{\partial x} \tag{1.10}$$

The first of these should be interpreted as 'multiply by x' and the second as 'differentiate with respect to x and multiply by $-i\hbar$'. We can easily verify that the wavefunction introduced in Section 1.2 to represent a particle with fixed momentum p is an eigenfunction of the operator $\hat{p}_x$ with eigenvalue p.

$$\hat{p}_x e^{ipx/\hbar} = -i\hbar\frac{\partial}{\partial x}e^{ipx/\hbar} = pe^{ipx/\hbar} \tag{1.11}$$

In classical physics all mechanical quantities of a one-dimensional system can be calculated if the two quantities x and p_x are known at the same time. In quantum mechanics the operators representing all of these mechanical quantities are constructed in exactly the same way from the operators $\hat{x}$ and $\hat{p}_x$. For example, consider the operator for the kinetic energy. In classical physics the kinetic energy T of a particle of mass m is equal to $\frac{1}{2}mv_x^2$, which can be expressed in terms of the momentum as $p_x^2/2m$. In quantum mechanics the kinetic energy is therefore represented by the operator

Position and momentum also occupy a special position in classical mechanics. If both are specified simultaneously then in principle it is possible to define the subsequent trajectory of the particle exactly.

$$\frac{\hat{p}_x^2}{2m} = -\frac{\hbar^2}{2m}\frac{\partial^2}{\partial x^2} \qquad (1.12)$$

Notice that the square of an operator is interpreted as the action of the operator twice in succession, hence the double derivative in the kinetic energy operator. Similarly the operator for the potential energy is $\hat{V}(x)$, 'multiply by the function $V(x)$'. The total energy is the sum of the kinetic energy and the potential energy. The total energy operator is therefore the sum of the kinetic energy operator and the potential energy operator, and is called the *Hamiltonian* $\hat{H}$.

$$\hat{H} = -\frac{\hbar^2}{2m}\frac{\partial^2}{\partial x^2} + \hat{V}(x) \qquad (1.13)$$

Writing the eigenvalue equation for the Hamiltonian operator we find

$$\hat{H}\psi = -\frac{\hbar^2}{2m}\frac{\partial^2\psi}{\partial x^2} + V(x)\psi = E\psi \qquad (1.14)$$

which is the Schrödinger equation. The eigenvalues of $\hat{H}$ are the allowed energy levels of the system.

Problem 1.3.1. *In the problem known as the particle in a box, a particle is confined to move on the x-axis between points $x = 0$ and $x = a$. The wavefunction for this particle must be zero outside these limits so that there is no probability of finding it outside the box. By continuity, it must therefore also be zero at the edges of the box. Inside the box the wavefunction obeys the Schrödinger equation with $V = 0$. Show by substitution into the Schrödinger equation that the functions $\psi_n = N\sin(n\pi x/a)$ are solutions, and that they are zero at the edges of the box only if n is an integer, hence find an expression for the corresponding energy levels. The first few wavefunctions are shown in Fig. 1.6.*

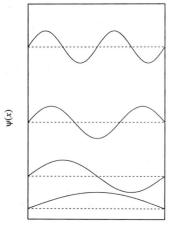

$x\,/\,a$

Fig. 1.6 Wavefunctions for the particle in a box. Each wavefunction is offset vertically by a distance proportional to its energy.

These ideas can be generalized to three dimensions. The momentum is now a vector quantity, and so the corresponding operator is the vector operator $-i\hbar\nabla$, whose operation on ψ gives the column vector:

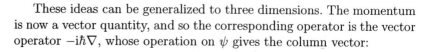

$$-i\hbar\nabla\psi = -i\hbar\left(\frac{\partial\psi}{\partial x}, \frac{\partial\psi}{\partial y}, \frac{\partial\psi}{\partial z}\right)^T \qquad (1.15)$$

(The superscript ^{T} indicates the transpose of the row vector.) Similarly the kinetic energy is given by Pythagoras' theorem:

$$T = \frac{1}{2m}(p_x^2 + p_y^2 + p_z^2) \qquad (1.16)$$

so that the corresponding operator is

$$\hat{T} = -\frac{\hbar^2}{2m}\nabla^2 = -\frac{\hbar^2}{2m}\left(\frac{\partial^2}{\partial x^2} + \frac{\partial^2}{\partial y^2} + \frac{\partial^2}{\partial z^2}\right) \qquad (1.17)$$

Another important example is the angular momentum operator. The classical angular momentum **l** is equal to the vector product **r** ∧ **p**, which is the vector

$$\mathbf{l} = \begin{vmatrix} \mathbf{i} & \mathbf{j} & \mathbf{k} \\ x & y & z \\ p_x & p_y & p_z \end{vmatrix} = \begin{pmatrix} yp_z - zp_y \\ zp_x - xp_z \\ xp_y - yp_x \end{pmatrix} \qquad (1.18)$$

where **i**, **j** and **k** are unit vectors in the x, y and z directions respectively. In classical systems with spherical symmetry the angular momentum is a 'constant of the motion', which means that it does not change unless an external influence is present. Such a property is also said to be *conserved*. The angular momentum is also conserved in a quantum system in the sense that wavefunctions with characteristic energies also have definite values of the angular momentum. Operators for the three components of the angular momentum are obtained if x, p_x etc. are replaced in eqn 1.18 by the corresponding operators. One of the most important physical properties of a spherical system is the magnitude of the angular momentum, whose square is given by $l^2 = l_x^2 + l_y^2 + l_z^2$. The operator for this quantity in quantum mechanics is therefore $\hat{l}^2 = \hat{l}_x^2 + \hat{l}_y^2 + \hat{l}_z^2$. It is sometimes convenient to transform the resulting operator into spherical polar coordinates, defined by

$$x = r \sin\theta \cos\phi, \ y = r \sin\theta \sin\phi, \ z = r\cos\theta \qquad (1.19)$$

r represents the distance of the point from the origin, θ is the angle of latitude measured from the z-axis, and ϕ is the longitude, measured from the xz-axial plane as the meridian (see Fig. 1.7). Following a certain amount of tedious algebra the result is obtained:

$$\hat{l}^2 = -\hbar^2 \left[\frac{1}{\sin^2\theta} \frac{\partial^2}{\partial\phi^2} + \frac{1}{\sin\theta} \frac{\partial}{\partial\theta}\left(\sin\theta \frac{\partial}{\partial\theta}\right) \right] \qquad (1.20)$$

The reader may notice that this is essentially the angular part of the ∇^2 operator expressed in spherical polar coordinates.[5]

Problem 1.3.3. *As a simpler example, consider a system with axial symmetry along the z-axis; the angular momentum of the motion about this axis is a constant of the motion. The simplest example of such a system is a particle confined to move in the xy-plane on a ring of radius a centred on the origin. The angular momentum is then directed along the z-axis. Find the angular momentum operator in cartesian coordinates, and then transform to polar coordinates ($x = r\cos\theta$ and $y = r\sin\theta$) to find $\hat{l}_z = -i\hbar\partial/\partial\theta$. [Hint: use the z component of the angular momentum from eqn 1.18, and treat the radial coordinate r as a constant.]*

Expectations

Suppose that an observable A is represented by an operator $\hat{A}$. If the wavefunction ψ is an eigenfunction of $\hat{A}$ with eigenvalue α ($\hat{A}\psi = \alpha\psi$), then the system has a definite value α of the observable A, and a mea-

Problem 1.3.2. *Write the Hamiltonian operator for the hydrogen atom, including the kinetic energy of the nucleus and the electron and the potential energy of interaction between the two particles.*

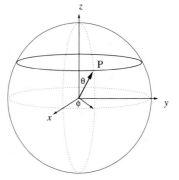

Fig. 1.7 The spherical polar coordinates. The point P lies on a sphere of radius r, a line of latitude which is defined by the angle θ measured from the z axis and a line of longitude ϕ, measured from the xz plane.

Problem 1.3.4. *Use the operator $\hat{l}_z$ derived in Problem 1.3.3 to show that the wavefunction $N\exp(il\theta)$ is an eigenfunction of the $\hat{l}_z$ operator with eigenvalue $l\hbar$. Find the normalization constant N.*

surement of A will yield the value α with probability one. However, if the wavefunction ψ is not an eigenfunction of $\hat{A}$ then a measurement of A may give any of the different possible eigenvalues of the operator $\hat{A}$ with different probabilities. We shall see in Section 1.5 how to calculate these probabilities; at this point we simply wish to calculate the average value of the observable A from the wavefunction. The mean or 'expectation' value of an observable A, denoted $< A >$, is defined by the integral

$$< A >= \int \psi^* \hat{A} \psi \, \mathrm{d}\tau \qquad (1.21)$$

In the special case where ψ is an eigenfunction of $\hat{A}$, of course, this integral gives the eigenvalue α:

$$< A >= \int \psi^* \hat{A} \psi \, \mathrm{d}\tau = \int \psi^* \alpha \psi \, \mathrm{d}\tau = \alpha \int \psi^* \psi \, \mathrm{d}\tau = \alpha \qquad (1.22)$$

[The first step follows because $\hat{A}\psi = \alpha\psi$; α can then be taken outside the integral because it is a constant; the final step follows because ψ is normalized (equation 1.8).]

The uncertainty in an observable A is defined as

$$\delta A = (< A^2 > - < A >^2)^{\frac{1}{2}} \qquad (1.23)$$

The uncertainty defined in this way is known in statistics as the standard deviation. It is the rms deviation from the mean.

It is clear from Problem 1.3.5 that the uncertainty in x can be calculated very simply from the wavefunction.

Problem 1.3.5. *Find the expectation of x and x^2 and hence the uncertainty in x for a particle in a box whose wavefunction is given in Problem 1.3.1.*

Dirac's bra-ket notation

Integrals of the type $\int \phi^* \hat{A} \psi \, \mathrm{d}\tau$, of which the expectation introduced above is a special case, arise frequently in applications. Dirac[6] proposed a shorthand representation which helps to distinguish wavefunctions, operators and constants from one another, and is less cumbersome than the standard notation of calculus.

A wavefunction ψ is represented by a symbol $|\psi >$. This symbol is called a *ket*. The character inside the ket is a label which should be sufficient to identify unambiguously the wavefunction of interest. For example, if the wavefunction is an eigenfunction of operator $\hat{A}$ then it may be identified by its eigenvalue, $|\alpha >$. Sometimes more than one label is needed to identify the wavefunction, for example three labels (quantum numbers) are needed to identify the electronic wavefunction of a hydrogen atom, $|nlm_l >$. The complex conjugate of the wavefunction, ψ^*, is similarly represented by a symbol $< \psi|$, which is called a *bra*. Integrals of the form $\int \phi^* \psi \, \mathrm{d}\tau$ and $\int \phi^* \hat{A} \psi \, \mathrm{d}\tau$ are represented, respectively, $< \phi|\psi >$ and $< \phi|\hat{A}|\psi >$, and are termed *bra(c)kets*. Note that when a bra and a ket are combined to form a bracket they are implicitly integrated over all available space.

We can now briefly summarise some of the definitions given above using the new notation.

1. A wavefunction $|\psi >$ is normalized if $< \psi|\psi >= 1$.

2. Two wavefunctions $|\psi>$ and $|\phi>$ are orthogonal if $<\phi|\psi>=0$.

3. A wavefunction $|\psi>$ is an eigenfunction of operator $\hat{A}$ with eigenvalue α if $\hat{A}|\psi>=\alpha|\psi>$.

4. The quantum mechanical expectation of an observable A is calculated from the wavefunction $|\psi>$ using the integral $<\psi|\hat{A}|\psi>$.

Problem 1.3.6. *Prove that* $<\phi|\psi>^*=<\psi|\phi>$, *and hence deduce that* $<\psi|\psi>$ *must be real and positive.*

Mathematical properties of operators

As stated above, all observable quantities are represented in quantum mechanics by operators. However, just as wavefunctions have to obey certain constraints, there are important restrictions on operators which represent observable quantities: linearity and hermiticity.

Linearity

An operator $\hat{A}$ is linear, if for any constants α and β, and wavefunctions $|f>$ and $|g>$

$$\hat{A}(\alpha|f>+\beta|g>) = \alpha\hat{A}|f>+\beta\hat{A}|g> \qquad (1.24)$$

As mentioned above, wavefunctions are frequently expressed as *superpositions* (linear combinations) of simpler wavefunctions, producing interference patterns. The linearity condition is necessary to ensure that the effect of applying an operator to each component of a superposition and then superposing the resulting waves is the same as applying the operator to the whole superposition at once.

Hermiticity

An operator $\hat{A}$ is hermitian if its integrals obey the relationship

$$<f|\hat{A}|g>^*=<g|\hat{A}|f> \qquad (1.25)$$

Hermitian operators have real eigenvalues, and are therefore suitable for representing observables. In addition their eigenfunctions are orthogonal

for all valid wavefunctions $|f>$ and $|g>$. This restriction ensures that eigenvalues of $\hat{A}$ are all real-valued (as opposed to complex). Since the eigenvalues of the operator are the only values that a measurement can take, they must evidently all be real.

The proof of this assertion is simple: let $|k>$ be an eigenfunction of the operator $\hat{A}$ with eigenvalue α_k. Then because $\hat{A}$ is hermitian it follows that $<k|\hat{A}|k>^*=<k|\hat{A}|k>$. The second of these is simply equal to α_k (see eqn 1.22), and so the first must be equal to α_k^*. If α_k is equal to its complex conjugate then it must be real. Hermitian operators have another useful property. Eigenfunctions of a hermitian operator belonging to different eigenvalues must be orthogonal. This can also be proved simply. The property of hermiticity implies that $<j|\hat{A}|k>^*=<k|\hat{A}|j>$. The second of these matrix elements (the right-hand side) is equal to $\alpha_j<k|j>$ and the left-hand side is $\alpha_k^*<j|k>^*$ which is also equal to $\alpha_k<k|j>$. Putting these together we find

$$(\alpha_j-\alpha_k')<k|j>=0 \qquad (1.26)$$

implying that if α_j and α_k are not equal then $< k|j >$ must be identically zero, or in other words that $|k >$ and $|j >$ are orthogonal, as claimed.

Problem 1.3.7. *Verify the following:*

(a) *The operator* $\hat{b}^+$ *(add the constant b) is not linear.*

(b) *The operator* $\hat{x}$ *(multiply by the real variable x) is both linear and hermitian.*

(c) *The operator* $\hat{\partial}_x$ *is linear but not hermitian—in fact it is anti-hermitian [integrate by parts].*

(d) *The operator* $\hat{p}_x$ *(multiply by* $-i\hbar$ *and differentiate with respect to x) is both linear and hermitian (the factor of* i *changes sign when the complex conjugate is taken).*

1.4   Commutation and uncertainty

A wavefunction with a characteristic, well-defined value of some observable quantity is an eigenfunction of the corresponding operator. However, such a wavefunction does not necessarily have a characteristic value of any other observable. For a wavefunction to have characteristic values of two observables simultaneously it is necessary for the corresponding operators to *commute*. Specifically this means that the action of the two operators (say $\hat{A}$ and $\hat{B}$) taken in succession on any wavefunction is identical to the action of the two operators taken in the reverse order,

> Operators are considered to operate to the right. Thus $\hat{A}\hat{B}$ means that the operator $\hat{B}$ is applied first and followed by $\hat{A}$:
> $\hat{A}\hat{B}|\psi > = \hat{A}(\hat{B}|\psi >)$

$$\hat{A}\hat{B} = \hat{B}\hat{A} \tag{1.27}$$

or equivalently the *commutator* of the two operators is zero

$$[\hat{A}, \hat{B}] = \hat{A}\hat{B} - \hat{B}\hat{A} = 0 \tag{1.28}$$

If two operators commute it is possible to find wavefunctions which have characteristic values of both observables, i.e. wavefunctions which are simultaneously eigenfunctions of both operators. If the eigenfunctions are not degenerate then all eigenfunctions of one operator must also be eigenfunctions of the other operator. This can be proved as follows: let $|\beta >$ be an eigenfunction of $\hat{B}$ with eigenvalue β, then

$$\hat{A}(\hat{B}|\beta >) = \beta\hat{A}|\beta > \tag{1.29}$$

but if the two operators commute,

$$\hat{A}(\hat{B}|\beta >) = \hat{B}(\hat{A}|\beta >) \tag{1.30}$$

and thus

$$\hat{B}(\hat{A}|\beta >) = \beta(\hat{A}|\beta >) \tag{1.31}$$

implying that $\hat{A}|\beta>$ is an eigenfunction of $\hat{B}$ with eigenvalue β. If there is only one such eigenfunction, $|\beta>$, then $\hat{A}|\beta>$ must be directly proportional to $|\beta>$. In other words $|\beta>$ is also an eigenfunction of $\hat{A}$. The converse of this result, that if two operators do not commute it is not in general possible to specify the values of the two corresponding observables simultaneously, is of the most fundamental importance, since it is the origin of the *uncertainty principle*. It is possible to show that the uncertainty product, which is the product of the standard deviations of the two noncommuting observables, must always be greater than half the expectation of the commutator of the two operators[7]:

$$\delta A \, \delta B \geq \frac{1}{2}| < \psi|[\hat{A}, \hat{B}]|\psi > | \qquad (1.32)$$

The most important specific example of the uncertainty principle is for the position and the momentum of a particle. Classically these are exactly the variables that need to be specified to define the subsequent trajectory of a particle, however their operators do not commute, as can be easily verified:

$$[\hat{x}, \hat{p}_x]\psi = -\mathrm{i}\hbar \left(x\frac{\mathrm{d}\psi}{\mathrm{d}x} - \frac{\mathrm{d}}{\mathrm{d}x}(x\psi) \right) = \mathrm{i}\hbar\psi \qquad (1.33)$$

which implies the uncertainty product

$$\delta x \delta p_x \geq \frac{\hbar}{2} \qquad (1.34)$$

Since it is not possible for a wavefunction to have characteristic values of two observables with noncommuting operators, it is not possible to specify exact simultaneous values for these observables. It should be noted that the uncertainty principle is rather deeper than the usual physical explanation might imply (the impossibility of measuring the position of a particle without disturbing its momentum). The uncertainty principle is actually inherent in the description of matter in terms of wavefunctions.

Problem 1.4.1. *(a) If n is an integer show that $[\hat{A}, \hat{A}^n] = 0$.*

(b) Show that $\hat{x}$ commutes with $\hat{p}_y$, and that $\hat{y}$ commutes with $\hat{p}_x$, but that $\hat{x}$ does not commute with $\hat{p}_x$.

(c) Prove that $[\hat{A}, \hat{B} + \hat{C}] = [\hat{A}, \hat{B}] + [\hat{A}, \hat{C}]$, and that $[\hat{A}, \hat{B}\hat{C}] = [\hat{A}, \hat{B}]\hat{C} + \hat{B}[\hat{A}, \hat{C}]$.

(d) Using the above result, and eqn 1.33, find the commutators $[\hat{x}, \hat{p}_x^2]$ and $[\hat{p}_x, \hat{x}^2]$.

(e) By repeated application of the results of problem (c), find the commutator $[\hat{x}\hat{p}_y, \hat{y}\hat{p}_x]$.

1.5   Expansion in eigenfunctions

Another important idea in quantum mechanics is the expansion in eigen-functions. As we already know, distinct eigenfunctions of a hermitian operator are mutually orthogonal (eqn 1.26). More importantly the set of eigenfunctions of an operator constitute what is known as a *complete set*, which means that any wavefunction for the system can be expressed as a superposition of them. To clarify this idea, suppose that operator $\hat{A}$ has eigenfunctions $|i>$ with eigenvalues α_i, i.e.

$$\hat{A}|i> = \alpha_i|i>\qquad(1.35)$$

Then any valid wavefunction for the system can be expressed as a superposition of the eigenfunctions:

$$\psi = \sum_i c_i|i>\qquad(1.36)$$

The complete set of eigenfunctions $|i>$ is frequently referred to as a *basis*. The wavefunction can be expressed in terms of the basis functions by specifying the coefficients c_i. Each eigenfunction $|i>$ can be thought of as a unit vector of an *orthonormal* coordinate system, because they are normalized, ($<i|i>=1$) and mutually orthogonal ($<i|j>=0$ if $i \neq j$). This analogy is exact: the wavefunction is represented in a particular basis by a vector, and the set of coefficients c_i is the expression of that vector in the coordinate system defined by the unit vectors $|i>$.

What is the significance of the coefficients c_i?

If ψ is not an eigenfunction of the operator $\hat{A}$, then a measurement of the quantity A may yield any of the different eigenvalues α_i. The significance of the coefficients c_i is that the probability of measuring the value α_i is equal to $c_i^*c_i$. This result is closely related to the Born interpretation of $\psi^*\psi$ as the spatial probability density of the system. We can make this result plausible by considering the expectation of A, defined in eqn 1.21 as $<A>=<\psi|\hat{A}|\psi>$. Substituting the expansion $\psi = \sum_i c_i|i>$, and the similar expansion $\psi^* = \sum_j c_j^* <j|$, we obtain

$$<A> = \sum_i \sum_j c_j^*c_i <j|\hat{A}|i>\qquad(1.37)$$

The formula can be simplified considerably since $|i>$ is an eigenfunction of $\hat{A}$, and distinct eigenfunctions of $\hat{A}$ are orthogonal.

$$<A> = \sum_i \sum_j c_j^*c_i\alpha_i <j|i> = \sum_i c_i^*c_i\alpha_i\qquad(1.38)$$

The sum over j disappears because the only nonzero term is the term with $j=i$. In statistics an expectation is a weighted average of all the possible outcomes, each value weighted by its probability.

$$< A >= \sum_i p_i \alpha_i \tag{1.39}$$

Comparing eqns 1.38 and 1.39 we can see that if the quantum mechanical expectation is to coincide with the statistical expectation, then $p_i = c_i^* c_i$. In other words the probability of measuring the value α_i is equal to $c_i^* c_i$. The argument given above is not rigorous, but can be made so by noting that the same formulas hold for the expectation of $\hat{A}^n$ if the quantity α_i in each summation is repaced by α_i^n. If all moments of a random variable calculated from two probability distributions are equal, then the two distributions are equal.[8]

How can the coefficients c_i be determined?

The key to calculating the coefficients c_i is the orthonormality of the eigenfunctions. In the vector representation the coefficient c_j is the component of the vector in the direction of one of the base unit vectors $|j>$, and it can therefore be found by taking the scalar product of the vector with $< j|$,

$$< j|\psi >= \sum_i c_i < j|i >= c_j \tag{1.40}$$

The second step follows because the basis functions are orthonormal, so that all terms in the summation are zero except the term with $i = j$. Thus in order to find the coefficients c_j it is necessary only to evaluate the integrals $< j|\psi >$.

1.6 The variational principle

The possibility of expansion in eigenfunctions has an important consequence, known as the variational principle. The essential content of this principle is the seemingly obvious statement that the expectation of a quantity must be greater than (or equal to) its lowest possible value, the smallest eigenvalue. This is particularly useful in chemistry where it is applied to the energy of a system. The proof of the variational principle is almost as simple as its statement. We start from the expression for the expectation of the quantity A from the previous section.

$$< A >= \sum_i c_i^* c_i \alpha_i \tag{1.41}$$

But since all the α_i are greater than (or equal to) the smallest eigenvalue α_0 and since every quantity $c_i^* c_i$ is positive, each term in this summation must be greater than or equal to a similar term in which α_i is replaced by α_0.

$$\sum_i c_i^* c_i \alpha_i \geq \sum_i c_i^* c_i \alpha_0 = \alpha_0 \sum_i c_i^* c_i = \alpha_0 \tag{1.42}$$

(The final step follows from normalization, see Problem 1.5.1).

This principle is important in.chemistry because although we often know the operator for a particular quantity, it is rare to be able to find either the eigenfunctions or the eigenvalues exactly. For example, we

Problem 1.5.1. *If $c_i^* c_i$ represents the probability of finding the value α_i in a measurement, then the sum of all these probabilities should equal one, $\sum_i c_i^* c_i = 1$. Prove that this is true by considering the normalization condition $< \psi|\psi >= 1$.*

Problem 1.5.2. *A particle on a ring is prepared with a wavefunction equal to $\frac{1}{\sqrt{\pi}}$ between $\theta = 0$ and $\theta = \pi$, and 0 between $\theta = \pi$ and $\theta = 2\pi$. If a measurement of the angular momentum is made, calculate the probability of finding a particular value $l\hbar$. [The angular momentum eigenfunctions are $|l >= (2\pi)^{-\frac{1}{2}} e^{il\theta}$].*

may know the exact Hamiltonian operator for a molecule, but we cannot solve the Schrödinger equation exactly for the ground state energy. Under these conditions it is necessary to resort to approximate forms for the eigenfunction. The variational principle guarantees that the expectation energy calculated from the approximate wavefunction will always be greater than the true ground state energy. We thus have an easy and unambiguous method for deciding which of two approximate wavefunctions is the best—it will be the approximation which gives the lowest energy. Similarly we have a strategy for finding the best approximation of a particular form—it will be the approximation which minimizes the energy. For example, if we introduce an approximate wavefunction which includes some variable parameter, then the optimum value of the parameter can be found by minimizing the expectation energy with respect to it. This application is worked through in more detail in II.

1.7 Transition probabilities

The most useful probes of the wavefunctions of atoms and molecules are the various forms of spectroscopy. In spectroscopy an atom or molecule starting with some wavefunction $|i>$ (usually an eigenfunction of the Hamiltonian), is probed with light (or some other particle). The light disturbs the molecule, altering the Hamiltonian while it interacts, finally leaving the molecule in another state. At this point we give only a brief insight into the quantum mechanics of the process. A more complete description will be given in II.

The action of the perturbation in causing a transition from the initial state $|i>$ to the final state $|f>$ can be represented by a hermitian operator $\hat{A}$, which operates on $|i>$ giving a superposition of other possible states:

$$|f> = \hat{A}|i> = \sum_j a_{ji}|j> \tag{1.43}$$

From the immediately preceding discussion we can see that a particular coefficient, a_{ki}, can be projected out of this superposition by taking a scalar product with $<k|$.

$$<k|\hat{A}|i> = a_{ki} \tag{1.44}$$

The set of coefficients a_{ki} for all possible k and i form a matrix, which represents the effect of the operator $\hat{A}$ on the basis functions. For this reason integrals of the form $<k|\hat{A}|i>$ are called *matrix elements*. The probability of the molecule being found in state $|k>$ at the end of the interaction is $|<k|\hat{A}|i>|^2$. This quantity is called a transition probability for obvious reasons. Of course in reality the transition probabilities will depend on the length of time for which the interaction is applied, but the basic result is always the same: that the transition rate is proportional to the square of the matrix element of an operator representing the effect of the interaction.

There is further discussion of transition probabilities, the intensities of spectral lines and selection rules, which determine when a transition is forbidden, in Chapters 2, 5 and in II.

In spectroscopy the oscillating electric field of the light interacts with the dipole moment of the molecule. Transitions are controlled by the dipole moment operator, $\hat{\mu}$.

References

1. Eisberg, R. M. (1961). *Fundamentals of modern physics*, p.81. Wiley, New York.
2. Feynman, R. P., Leighton, R. B. and Sands, M. (1965). *The Feynman lectures on physics*, Vol. I, Ch. 29 and Vol. III, Ch. 1. Addison-Wesley, Reading MA.
3. Schiff, L. I. (1968). *Quantum mechanics*, 3rd edn., Ch. 6. McGraw-Hill, Tokyo.
4. Feynman, R. P. and Hibbs, A. R. (1965). *Quantum mechanics and path integrals*. McGraw-Hill, New York.
5. Stephenson, G. (1973). *Mathematical methods for science students*, p. 362. Longman, London.
6. Dirac, P. A. M. (1958). *Principles of quantum mechanics*, 4th edn., Ch. 1. Oxford University Press, Oxford.
7. Landau, L. D. and Lifschitz, E. M. (1977). *Quantum mechanics*, 1977, 3rd edn., p. 47. Pergamon Press, Oxford.
8. Moran, P. A. P. (1968). *An introduction to probability theory*, Section 6.4, p. 259. Oxford University Press, Oxford.

2 Separations

2.1 Multi-dimensional problems in chemistry

In Chapter 1 we discussed the way in which observable quantities are represented in quantum mechanics by operators. The most important of these operators is the Hamiltonian operator, which represents the total energy of the system. In some simple systems the Hamiltonian can be expressed in terms of a single variable, and the resulting eigenfunctions are functions of that variable alone. For example the Schrödinger equation for a simple harmonic oscillator can be written in terms of the single variable x:

$$-\frac{\hbar^2}{2m}\frac{\partial^2\psi}{\partial x^2} + \frac{1}{2}kx^2\psi = E\psi \tag{2.1}$$

As another example, the Schrödinger equation for a particle confined to a ring of radius a can be transformed into polar coordinates (see Problem 1.3.3); the wavefunction depends only on the angle θ

$$-\frac{\hbar^2}{2ma^2}\frac{\partial^2\psi}{\partial\theta^2} = E\psi \tag{2.2}$$

However, these single-variable problems are the exception rather than the rule. Even in the hydrogen atom three coordinates are required to specify the position of the electron relative to the nucleus, and so the electronic wavefunctions depend on three coordinates. The Schrödinger equation becomes a three-dimensional partial differential equation (see Problem 1.3.2). In larger atoms the electronic wavefunctions depend on three coordinates for each electron in the atom, and the dimensionality of the Schrödinger equation consequently becomes much larger.

Although many mathematical methods are available for one-dimensional eigenvalue problems, multi-dimensional partial differential equations are much more difficult to solve, often requiring computational methods. The first step in the strategy for dealing with multi-dimensional Schrödinger equations is therefore to search for some means of reducing the problem to a collection of one-dimensional equations, which can be solved separately. The procedure involved is called the *separation of variables*. Sometimes (as for the hydrogen atom) this can be done exactly, but sometimes the separation is only approximate, and the resulting solutions are therefore also only approximate. In this chapter we first outline the method of separation of variables, and we then consider how it is applied to problems in chemistry, both where it is exact, and where it is not.

2.2 The separation of variables

In the development of the method of separation of variables we shall initially concentrate on wavefunctions which depend only on two variables, x and y. We wish to find the eigenfunctions of the Hamiltonian operator, subject to suitable boundary conditions, and the corresponding eigenvalues are the allowed energy levels. The method consists in searching for a solution of the form $\psi = X(x)Y(y)$, i.e. a product of a function of variable x alone (X) and a function of y alone (Y). If solutions of this form, obeying the boundary conditions, can be found then the problem is solved.

Separation of variables is a mathematical method for reducing a problem with many variables to a set of uncoupled equations, each of one variable

As an illustration of the method we consider the example of a particle confined inside a two-dimensional square box of side a. The potential energy of the particle in the box is zero. In classical mechanics the energy of the particle is $(p_x^2 + p_y^2)/2m$, and so the Schrödinger equation is given by

$$-\frac{\hbar^2}{2m}\left(\frac{\partial^2 \psi}{\partial x^2} + \frac{\partial^2 \psi}{\partial y^2}\right) = E\psi \qquad (2.3)$$

with the boundary conditions that ψ must vanish at the edges of the box: i.e. at $x = 0, y = 0, x = a$ and $y = a$. (ψ is zero outside the box and because ψ must be continuous, it must therefore also be zero at the edges of the box.)

We now assume that the solution is of the form $\psi(x, y) = X(x)Y(y)$. Substituting this form into the Schrödinger equation we obtain:

$$-\frac{\hbar^2}{2m}\left(Y\frac{d^2 X}{dx^2} + X\frac{d^2 Y}{dy^2}\right) = EXY \qquad (2.4)$$

Dividing both sides by XY this becomes

$$-\frac{\hbar^2}{2m}\left(\frac{1}{X}\frac{d^2 X}{dx^2} + \frac{1}{Y}\frac{d^2 Y}{dy^2}\right) = E \qquad (2.5)$$

The first term depends only on the variable x, and the second term depends only on y, but E is a constant, independent of both x and y, implying that the first and second terms must separately be constant. We can understand this conclusion by considering what happens to the equation when x and y vary. Since x and y are independent variables they can be varied at will. First consider what happens if x is changed but y is held constant. The second term must be constant since it depends only on y, and y is being held constant. If both E and the second term are constant then the first term itself must also be constant even though x is being varied; let us call it E_1. An exactly reciprocal argument in which x is held constant but y is varied permits us to conclude that the second term is also constant, independent of y, which we denote E_2. The two *separation constants*, E_1 and E_2, must add up to E for the equation to hold. The second-order equation is thus separated to the pair of equations

$$-\frac{\hbar^2}{2m}\frac{d^2X}{dx^2} = E_1 X \text{ and } -\frac{\hbar^2}{2m}\frac{d^2Y}{dy^2} = E_2 Y \tag{2.6}$$

These equations are separate one-dimensional differential equations for the two functions X and Y. With the boundary conditions given, each of these equations can be recognized as the Schrödinger equation for a particle in a one-dimensional box, whose solutions are already known from Problem 1.3.1. Of course, the final energy is $E = E_1 + E_2$.

2.3 Spherical symmetry

Many real systems have spherical symmetry. The strategy is to transform to spherical polar coordinates and then separate the three variables r, θ and ϕ.

Many important problems of chemical significance can be expressed in terms of a simple spherically symmetrical potential energy. For example the potential energy of electrostatic attraction between the electron and the nucleus in a hydrogen atom depends only on the distance between the electron and the nucleus. Another example is the energy of a diatomic molecule, which does not depend on the orientation of the molecule in space. We therefore need to consider the motion of a particle in a spherical potential. The separation of variables is more difficult in this case than the simple separation described in Section 2.2, but because of the importance of spherical symmetry in chemistry we shall work through the separation in detail.

The Schrödinger equation for a spherical system is

$$-\frac{\hbar^2}{2m}\nabla^2\psi + V(r)\psi = E\psi \tag{2.7}$$

where the spherical symmetry means that the potential energy function $V(r)$ depends only on the variable r, representing the distance to the origin. Using the standard expansion of the operator ∇^2 in spherical polar coordinates[1] this can be written in the rather fearsome form

$$-\frac{\hbar^2}{2m}\left\{\frac{1}{r^2}\frac{\partial}{\partial r}\left(r^2\frac{\partial\psi}{\partial r}\right) + \frac{1}{r^2}\left[\frac{1}{\sin\theta}\frac{\partial}{\partial\theta}\left(\sin\theta\frac{\partial\psi}{\partial\theta}\right) + \frac{1}{\sin^2\theta}\frac{\partial^2\psi}{\partial\phi^2}\right]\right\}$$
$$+ V(r)\psi = E\psi \tag{2.8}$$

The part of this expression in square brackets, containing the angular derivatives (with respect to θ and ϕ) is the operator for the square of the angular momentum, which was introduced in eqn 1.20. Thus we have the more economical form

$$-\frac{\hbar^2}{2m}\frac{1}{r^2}\frac{\partial}{\partial r}\left(r^2\frac{\partial\psi}{\partial r}\right) + \frac{1}{2mr^2}\hat{l}^2\psi + V(r)\psi = E\psi \tag{2.9}$$

This equation is a partial differential equation in three variables, and to make any headway we need to try to separate the three variables. Since we have recognized the presence of the angular momentum operator $\hat{l}^2$ it makes sense to try to separate the variable r from the angular variables first, and then to solve the angular equation for the eigenfunctions of the angular momentum. We therefore search for a solution of the form

$R(r)Y(\theta, \phi)$. Multiplying both sides of the Schrödinger equation by r^2, applying the assumed form $\psi = RY$, dividing through by RY and rearranging, we obtain

$$-\frac{\hbar^2}{2m}\frac{1}{R}\frac{d}{dr}\left(r^2\frac{dR}{dr}\right) + (V(r) - E)r^2 = -\frac{1}{2mY}\hat{l}^2Y \tag{2.10}$$

The left-hand side clearly depends only on the variable r, whilst the right-hand side depends only on the angles θ and ϕ. Since r, θ and ϕ are independent variables, the RHS and the LHS must independently be equal to the same constant, which we denote $-W$. Hence we arrive at the separated equations:

$$-\frac{\hbar^2}{2m}\frac{1}{r^2}\frac{d}{dr}\left(r^2\frac{dR}{dr}\right) + \left(V(r) + \frac{W}{r^2}\right)R = ER \tag{2.11}$$

and

$$\hat{l}^2Y = 2mWY \tag{2.12}$$

Initially we concentrate on the second of these equations, which can be written explicitly

$$\hbar^2\left[\frac{1}{\sin\theta}\frac{\partial}{\partial\theta}\left(\sin\theta\frac{\partial Y}{\partial\theta}\right) + \frac{1}{\sin^2\theta}\frac{\partial^2 Y}{\partial\phi^2}\right] = -2mWY \tag{2.13}$$

The solutions are the eigenfunctions of the $\hat{l}^2$ operator, whose eigenvalues are the possible values of $2mW$. Once again, we search for solutions in which the two variables θ and ϕ are separated, i.e. solutions of the form $\Theta(\theta)\Phi(\phi)$, leading to the separation

$$-\frac{\sin\theta}{\Theta}\frac{d}{d\theta}\left(\sin\theta\frac{d\Theta}{d\theta}\right) - \frac{2mW}{\hbar^2}\sin^2\theta = \frac{1}{\Phi}\frac{d^2\Phi}{d\phi^2} \tag{2.14}$$

The left-hand side of the equation depends only on θ and the right-hand side only on ϕ. Each side of the equation must therefore be equal to the same constant, which we denote $-m_l^2$. The equation for Φ therefore becomes

$$\frac{d^2\Phi}{d\phi^2} = -m_l^2\Phi \tag{2.15}$$

which is the same as eqn 2.2 for the motion of a particle round a ring. This equation can be solved by elementary methods, giving $\Phi = N\exp[\pm im_l\phi]$. However, there are restrictions on the possible values of the constant m_l, because Φ must be a single-valued function. Since passing round one complete cycle from ϕ to $\phi + 2\pi$ brings us back to the same point on the ring again, $\Phi(\phi)$ must be equal to $\Phi(\phi + 2\pi)$. The only possible values of m_l are therefore positive and negative integers and zero, so that $\exp(2\pi im_l) = 1$. The function $\exp(im_l\phi)$ is also an eigenfunction of the $\hat{l}_z$ operator, with eigenvalue $\hbar m_l$.

$$\hat{l}_z e^{\pm i m_l \phi} = -i\hbar \frac{\partial}{\partial \phi} e^{\pm i m_l \phi} = \pm \hbar m_l e^{\pm i m_l \phi} \qquad (2.16)$$

$\pm \hbar m_l$ is therefore the z-component of the angular momentum.

The separated equation for the function Θ

$$\sin \theta \frac{d}{d\theta} \left(\sin \theta \frac{d\Theta}{d\theta} \right) = \left(-\frac{2mW}{\hbar^2} \sin^2 \theta + m_l^2 \right) \Theta \qquad (2.17)$$

is more difficult to solve. We present its solution in Section 3.4. The most important feature of the solution is the identification of W with the eigenvalue for the square of the angular momentum, $2mW = \hbar^2 l(l+1)$, where l is an integer constrained to be greater than or equal to $|m_l|$. We shall return to this in more detail in the next chapter.

The radial function therefore obeys the equation

$$-\frac{\hbar^2}{2mr^2} \frac{d}{dr} \left(r^2 \frac{dR}{dr} \right) + \left(V(r) + \frac{\hbar^2 l(l+1)}{2mr^2} \right) R = ER \qquad (2.18)$$

The detailed solution to the equation will depend on the form of the potential energy $V(r)$ for the particular case under investigation. However, the equation can be reduced to a slightly simpler form by the substitution $S(r) = rR(r)$, in which case we find

$$-\frac{\hbar^2}{2m} \frac{d^2 S}{dr^2} + \left(V(r) + \frac{\hbar^2 l(l+1)}{2mr^2} \right) S = ES \qquad (2.19)$$

which is the Schrödinger equation for a one-dimensional system with a potential energy $(V(r) + \hbar^2 l(l+1)/2mr^2)$. The second term is a repulsive potential pushing the particle out from the origin, resulting from the fact that the radial equation is solved in a rotating frame of reference and is often known as the *centrifugal energy*. It has important consequences for the detailed understanding of chemical reactions and of the energy levels of molecules near to dissociation.[2] Figure 2.1 shows a schematic potential energy curve for a diatomic molecule close to its dissociation limit, showing the development of a centrifugal barrier to atom recombination as the angular momentum of the system increases.

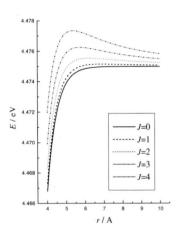

Fig. 2.1 Potential energy curve for H_2 at long bond extensions, showing the development of a centrifugal barrier as angular momentum increases, $J = 0, 1, 2, 3, 4$.

Six coordinates are required to specify the positions of two particles in space. The strategy for dealing with a system of two interacting particles is to transform to centre of mass coordinates and relative coordinates. The motion can then be separated into a translational motion of the centre of mass and an internal motion of the relative coordinate.

2.4 Two interacting particles

Another important exact separation of variables is illustrated by the hydrogen atom. The energy can be expressed as the sum of three terms, the kinetic energy of the nucleus, the kinetic energy of the electron and the potential energy of the electrostatic electron–nucleus interaction. The Schrödinger equation is therefore (see Problem 1.3.2)

$$-\frac{\hbar^2}{2m_n} \nabla_n^2 \psi - \frac{\hbar^2}{2m_e} \nabla_e^2 \psi - \frac{e^2}{4\pi \epsilon_o r} \psi = E\psi \qquad (2.20)$$

where ∇_n^2 denotes differentiation with respect to the nucleus coordinates and ∇_e^2 with respect to the electron coordinates. In the corresponding classical system the relative motion of the electron with respect to the nucleus is independent of the motion of the centre of mass. If we introduce new coordinates in which the subscript r denotes the coordinates of the electron relative to the position of the nucleus, and the subscript c denotes the centre of mass coordinates, i.e.

$$\mathbf{r}_r = \mathbf{r}_e - \mathbf{r}_n \text{ and } \mathbf{r}_c = \frac{m_n\mathbf{r}_n + m_e\mathbf{r}_e}{m_e + m_n} \qquad (2.21)$$

then the Schrödinger equation becomes

$$-\frac{\hbar^2}{2(m_n + m_e)}\nabla_c^2\psi - \frac{\hbar^2}{2\mu}\nabla_r^2\psi - \frac{e^2}{4\pi\epsilon_o r}\psi = E\psi \qquad (2.22)$$

where $\mu = m_e m_n/(m_e + m_n)$ and is known as the reduced mass. The wavefunction can be separated in the form $\chi(\mathbf{r}_c)R(\mathbf{r}_r)$, giving

$$-\frac{\hbar^2}{2(m_n + m_e)}\nabla_c^2\chi = E_c\chi \text{ and } -\frac{\hbar^2}{2\mu}\nabla_r^2 R - \frac{e^2}{4\pi\epsilon_o r}R = E_r R \qquad (2.23)$$

The former equation describes a particle freely moving in space, which we shall solve in Section 3.1. The latter equation is spherically symmetrical, and can be further simplified using the separation of variables described in Section 2.3.

The total energy E is equal to $E_c + E_r$. E_r is often described as the electronic energy of the atom, but the relative coordinate is partly composed of the motion of the nucleus about the mutual centre of mass. If the nucleus is treated as stationary, the same equation is found for the electronic motion, except that the reduced mass μ is replaced by the electron mass m_e. The resulting energy levels will have a relative error of $m_e/(m_e + m_n)$, or about one part in 2000. This degree of accuracy is acceptable for all but the most precise work, and so in analysing electronic motions it is usually acceptable to treat the nuclei as stationary. This approximation is the subject of the next section.

2.5   The Born–Oppenheimer separation

The physical reason behind the very small error incurred by neglecting the contribution of the nuclear motion to the internal energy of the hydrogen atom is that the mass of the nucleus is very large compared to the mass of the electron. The atomic centre of mass is therefore very close to the nucleus and the internal motion is composed almost entirely of the electronic motion. In addition the electron moves much faster than the nucleus relative to the centre of mass. A similar observation holds for molecules, and to a good approximation, originally due to Born and Oppenheimer, the total wavefunction of a molecule, which properly depends on the coordinates of the nuclei and the electrons, but can be separated approximately into a product of an electronic wavefunction and a nuclear wavefunction. The details are complicated, and we con-

Atomic units

It is usual in quantum chemistry to express all quantities in so-called *atomic units*. The unit of length is taken to be the Bohr radius, $a_0 = 4\pi\epsilon_0\hbar^2/me^2$ in which m and e are the mass and charge on an electron. The unit of energy (the *Hartree*) is taken to be $me^4/(4\pi\epsilon_0)^2\hbar^2$.

Problem 2.4.1. *Verify that a_0 has the dimensions of length, and that the Hartree has the dimensions of energy.*

Problem 2.4.2. *Transform eqn 2.23 into atomic units to obtain*

$$-\frac{m}{2\mu}\nabla^2\psi - \frac{1}{r}\psi = E\psi \qquad (2.24)$$

If the difference between m and μ is neglected, the factor multiplying the ∇^2 operator becomes $-\frac{1}{2}$.

Nearly all chemical problems are analysed using the Born–Oppenheimer approximation, in which electronic and nuclear motions are separated because electrons move much faster than nuclei.

centrate on the results, making no attempt to present derivations. The interested and brave reader is referred to the treatment by Bunker.[3]

The full Schrödinger equation for a molecule is given by

$$-\frac{\hbar^2}{2}\sum_n \frac{1}{M_n}\nabla_n^2\psi - \frac{\hbar^2}{2m_e}\sum_i \nabla_i^2\psi - \sum_{i,n}\frac{Z_n e^2}{4\pi\epsilon_o r_{in}}\psi$$

$$+\sum_{i>j}\frac{e^2}{4\pi\epsilon_o r_{ij}}\psi + \sum_{m>n}\frac{Z_m Z_n e^2}{4\pi\epsilon_o r_{mn}}\psi = E\psi \qquad (2.25)$$

where the subscripts e, i and j refer to the electrons and the subscripts m and n refer to the nuclei. Although this equation looks rather fearsome it is simple to see the origin of each term. The first term represents the kinetic energy of the nuclei (cf. eqn 1.17), and the second term that of the electrons; the remaining three terms comprise the potential energy of electrostatic attraction between electrons and nuclei, the electron–electron repulsion and the nucleus–nucleus repulsion. This equation may be simple to set up, but solving it is a different matter.

As with the H atom the first stage is to refer all coordinates to the molecular centre of mass, separating out the motion of the centre of mass of the molecule, which behaves like a free particle. However, this transformation also has the unfortunate side effect of introducing an extra term into the Schrödinger equation[3], which represents an interaction of the momentum of each particle in the molecule with all the other particles; in particular it introduces a coupling of nuclear and electronic motions, which means that they cannot strictly be separated. Although the three centre of mass coordinates have been separated out, reducing the dimensionality of the Schrödinger equation by three, the remaining equation is still far too complicated, and at this point we make the Born–Oppenheimer approximation.

In the Born–Oppenheimer approximation the electronic motion is analysed assuming the nuclei to be stationary. This gives an electronic energy $V_e(\mathbf{r}_n)$ which depends on the positions of the stationary nuclei.

Since the nuclei move much more slowly than the electrons relative to the molecular centre of mass it is assumed that we can find a wavefunction for the electrons alone (ψ_e) under the approximation that the nuclei are stationary. All terms in the Hamiltonian describing the motion of the nuclear framework are therefore neglected, leading to separation of the electronic and nuclear coordinates. The terms in the summation discussed above which couple the electron momenta are of the same order of magnitude, and are also neglected. The resulting Schrödinger equation for the electronic wavefunction with a fixed set of nuclear coordinates is

$$-\frac{\hbar^2}{2m_e}\sum_i \nabla_e^2\psi_e - \sum_{i,n}\frac{Z_n e^2}{4\pi\epsilon_o r_{in}}\psi_e + \sum_{i>j}\frac{e^2}{4\pi\epsilon_o r_{ij}}\psi_e = V_e\psi_e \qquad (2.26)$$

with a separation constant V_e which represents the electronic energy of the molecule with the nuclei in their fixed positions. The equation is solved by computational methods, to give a set of electronic wavefunctions which are functions of the electron coordinates and a corresponding set of eigenvalues for the electronic energy of the molecule. Each of these energy levels corresponds to a different electronic state of the molecule. In an atom there is no internal nuclear motion to consider, and the spectrum of the atom may be interpreted in terms of these quantum states. However, for a molecule a different set of energy levels will be obtained for every possible set of nuclear coordinates. Thus, in the Born–Oppenheimer approximation, the electronic energy of each state of the molecule is a function of the nuclear coordinates $\mathbf{r}_n$; it therefore acts as a potential energy for the nuclear motions, $V_e(\mathbf{r}_n)$. The energy required to move the nuclei from one configuration to another is the sum of the change in electronic energy, V_e and the nucleus–nucleus repulsion energy, $V_n(\mathbf{r}_n)$.

The idea of a potential energy curve or surface is central to many aspects of chemistry, for example in the unravelling of molecular spectra, in analysing the forces between molecules or in interpreting the dynamics of a chemical reaction. The potential energy curves for the first few electronic states of the O_2 molecule are shown in Fig. 2.2. Crudely speaking, each state differs in the way its electrons are distributed in the available orbitals.

The next stage of the problem is to solve the Schrödinger equation for the nuclear wavefunction. To do this we need to fix our attention on a particular electronic state since the potential energy function will be different for every electronic state. Having chosen a particular state and its corresponding potential energy function $V_e + V_n$ we obtain the equation

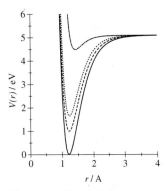

Fig. 2.2 Potential energy curves for the lowest four states of the O_2 molecule. The four states are designated $X\,^3\Sigma_g^-$, $a\,^1\Delta_g$, $b\,^1\Sigma_g^+$ and $A\,^3\Sigma_u^+$.

$$-\frac{\hbar^2}{2}\sum_i \frac{1}{M_i}\nabla_i^2\psi_n + \frac{\hbar^2}{2M}\sum_{i,j}\nabla_i\cdot\nabla_j\psi_n + (V_e + V_n)\psi_n = E\psi_n \quad (2.27)$$

where the summations run over the nuclei only. The eigenfunctions ψ_n describe the motion of the nuclei over the potential energy surface, and the eigenvalues E represent the total energy of the molecule (electronic and nuclear).

In summary, the Born–Oppenheimer approximation simplifies the molecular Schrödinger equation by separating the electronic motion from the nuclear motion. The justification for this separation is that the electrons move much faster than the nuclei with respect to the centre of mass and hence the nuclei can be thought of as seeing only an average electron distribution, whereas the electrons can react essentially instantaneously to any change in the nuclear configuration. The electronic equation is separated out under the approximation that the nuclei are stationary, and gives an electronic wavefunction $\psi_e(\mathbf{r}_e, \mathbf{r}_n)$, which depends both on the electron coordinates $\mathbf{r}_e$ and on the fixed nuclear coordinates $\mathbf{r}_n$. The corresponding electronic energy levels V_e also depend on $\mathbf{r}_n$. The combination of V_e with the nuclear repulsion energy V_n gives a potential energy surface, which is used for solving the nuclear Schrödinger equation. The solutions of this equation $\psi_n(\mathbf{r}_n)$ describe the nuclear motions of the molecule (rotation and vibration) and the eigenvalues E are the permitted energy levels of the molecule (including electronic and nuclear contributions). Finally, the overall molecular wavefunction is assembled by combining the electronic and nuclear wavefunctions:

$$\psi(\mathbf{r}_e, \mathbf{r}_n) = \psi_e(\mathbf{r}_e, \mathbf{r}_n)\psi_n(\mathbf{r}_n) \qquad (2.28)$$

The Franck–Condon principle

An electronic transition is assumed to occur so rapidly that the nuclei do not have time to respond. Effectively the potential energy curve for the molecule changes instantaneously. The nuclei find themselves with the wrong vibrational wavefunction and vibrations are excited.

One important consequence of the Born–Oppenheimer separation in spectroscopy is the Franck–Condon principle, which governs the intensities of the vibrational fine structure of an electronic transition in a molecular spectrum. As we have already seen in Section 1.7 the intensity of a transition is proportional to the square of the dipole moment matrix element. Thus the rate of transition from an initial state $|i>$ to a final state $|f>$ is proportional to $|<f|\hat{\mu}|i>|^2$.

The dipole moment of a molecule is a measure of its charge distribution, and the dipole moment operator is equal to the vector sum of the position vectors of all the particles in the molecule, weighted by their charges:

$$\hat{\mu} = \sum_j z_j \mathbf{r}_j \qquad (2.29)$$

This operator can be split into two parts, one of which contains the coordinates of the nuclei, and the other the coordinates of the electrons. The dipole moment matrix element which controls the intensity of a transition from state $|e''n''>$ to state $|e'n'>$ is now of the form

$$\mu_{e''n'',e'n'} = <e'n'|\hat{\mu}_e + \hat{\mu}_n|e''n''>$$

When we make the Born–Oppenheimer separation $|en> = |e>|n>$ and assume that the electronic wavefunctions depend only weakly on the nuclear coordinates, this matrix element separates into two parts:

$$\mu_{e''n'',e'n'} = <e'|\hat{\mu}_e|e''><n'|n''> + <e'|e''><n'|\hat{\mu}_n|n''> \qquad (2.30)$$

The second term is zero in an electronic transition because the electronic wavefunction changes from a state with one energy to a state with a different energy for almost all nuclear configurations. The two electronic wavefunctions are therefore orthogonal (see eqn 1.26). The first term, however, is not zero because although the vibrational quantum numbers for states $|n''>$ and $|n'>$ may be different, the two vibrational wavefunctions are from different electronic states, with different potential energy functions, and so they are solutions to different Schrödinger equations. There is therefore no reason for them to be orthogonal, except perhaps by accident. The matrix element is therefore the product of two parts: an electronic part $<e'|\hat{\mu}_e|e''>$, which contains the electronic dipole moment operator and is constant for a given electronic transition, and a nuclear part $<n'|n''>$, which is the overlap integral between the initial and final vibrational wavefunctions. The first term determines whether the transition is electronically allowed; if it is identically zero the transition has zero intensity and is called a *forbidden transition* (see Chapter 5). The second term shows that the intensity of the vibrational structure within an electronic transition will be proportional to the square of the overlap integral, $| < n'|n'' > |^2$. This term is called the *Franck–Condon factor*.

An alternative approach to the Franck–Condon principle will be described in II.

2.6 Spin

The reader may have noticed that we have not yet mentioned the idea of spin. The reason for this is that spin has no natural place in the development of Schrödinger's theory. It arises from a more detailed theory due to Dirac, which incorporates time in a relativistically consistent manner, but is beyond the scope of this book. If we are dealing only with light atoms and molecules made up of them, relativistic corrections are small and to a first approximation spin does not contribute to the energy. However, spin plays an extremely important role in chemistry—the structure of the periodic table arises because of the Pauli exclusion principle, which is a spin effect, and there are other important symmetry consequences of spin, which we discuss in II.

Rather than adopting the full relativistic theory of Dirac, spin is *patched into* the Schrödinger theory; the full wavefunction for a single particle is assumed to be of the form $\psi(\mathbf{r}) \times \chi_s(\omega)$. The first term is the *orbital* wavefunction, which depends on the spatial coordinates of the particle. The second term is the spin wavefunction, which depends on a 'spin coordinate' ω. The two parts multiplied together in this way are frequently referred to as a *spin orbital*, particularly if the particle described is an electron. We can write the full wavefunction in this separated form because in a nonrelativistic theory there is no interaction between the spin and the spatial motion of the particle. One consequence of this separation arises in electronic spectroscopy. We have already seen in the previous section that the intensity of an electronic transition depends on the matrix element $< e'|\hat{\mu}_e|e'' >$. If the spin is now included we can separate this matrix element into the product $< e'|\hat{\mu}_e|e'' >< s'|s'' >$, in which $|e >$ now refers to the orbital function and $|s >$ to the spin function. The matrix element $< s'|s'' >$ is zero unless $s' = s''$ because spin functions with different spin quantum numbers are orthogonal. We conclude therefore that electronic transitions in which the spin wavefunction changes are forbidden.

In the remainder of this chapter we describe other approximate separations which are often applied in solving the two separated parts of the molecular Schrödinger equation.

2.7 The orbital approximation

Once the electronic and nuclear coordinates have been separated, as described in Section 2.5 the next stage is to solve the electronic eqn 2.26.

In the orbital approximation each electron is assigned to

an orbital, these one-electron wavefunctions are multiplied together to form the overall electronic wavefunction. This procedure effectively assumes that the electrons are independent of one another, by averaging out the electron repulsions.

This must be done first because it provides the potential energy function for the solution of the nuclear eqn 2.27. The electrons interact through the electron–electron repulsion term:

$$\sum_{i>j} \frac{e^2}{4\pi\epsilon_o r_{ij}} \psi_e$$

If this term is neglected then the remainder of the Hamiltonian $\hat{H}_o$ can be written in the form

$$\hat{H}_o = \sum_i \hat{h}_i$$

where each $\hat{h}_i$ is a mini-Hamiltonian depending on the coordinates of one electron alone,

$$\hat{h}_i = -\frac{\hbar^2}{2m_e}\nabla_i^2 - \sum_n \frac{Z_n e^2}{4\pi\epsilon_o r_{in}} \qquad (2.31)$$

and containing the kinetic energy operator for the electron and the potential energy of attraction with the static nuclear framework. Thus if the electron repulsions are neglected, the coordinates of each electron can be separated out in turn and a solution of the modified equation can be found as a product of single-electron wavefunctions:

$$\psi_e = \phi_1(\mathbf{r}_1)\phi_2(\mathbf{r}_2)\dots\phi_n(\mathbf{r}_n) \qquad (2.32)$$

Neglecting the electron–electron repulsion is a serious approximation, and introduces gross errors, however, the picture implied by a solution of this form is far too useful to be easily discarded. In this approximation each electron is described independently by its own wavefunction, and the overall electronic wavefunction of the atom or molecule can be built up from the set of one-electron wavefunctions. This approximation is actually the way that chemists usually think of the electronic wavefunction: each one-electron wavefunction is an *orbital* and each electron is assigned to an orbital. The approximation that the electronic wavefunction can be expressed as a product of orbitals, and that therefore the electronic energy is the sum of the individual orbital energies, is known as the *orbital approximation*. It is important to realize that the idea of an orbital, and of an electron configuration in which each electron is assigned to an orbital, which is so vital in many areas of chemistry, is in fact an approximation which neglects any coupling between the electrons. Neither orbitals, nor electron configurations actually exist, except in truly one-electron systems.

Self-consistent field theory is an iterative method used to calculate the orbitals one by one. Each one-electron wavefunction is solved with a repulsion energy averaged over all the other electrons.

Self-consistent field theory

The quantitative picture is not as bad as it may seem from the previous discussion. The wavefunctions predicted by the orbital approximation have the correct symmetry, and so they are very useful for classifying the symmetries of the true electronic wavefunctions, or states of the molecule. Furthermore the orbital approximation can be improved significantly by using the *self-consistent field* method, which includes an average repulsive potential energy for each electron, while retaining the product form of the wavefunction.

The method can be illustrated by considering the contribution to the Schrödinger equation for electron 1 from interaction with electron 2. In the SCF method this quantity is averaged over the coordinates of electron 2 to give a contribution to the effective potential energy $v_{12}(\mathbf{r}_1)$.

$$v_{12}(\mathbf{r}_1) = \frac{e^2}{4\pi\epsilon_o} \int \frac{\phi_2^*\phi_2}{r_{12}} d\tau_2 \tag{2.33}$$

All dependence on $\mathbf{r}_2$ has been integrated out to give an average potential energy of repulsion. Thus the one-electron Schrödinger equation for electron 1 becomes

$$\hat{h}_1\phi_1 + \sum_j v_{1j}(\mathbf{r}_1)\phi_1 = E_1\phi_1 \tag{2.34}$$

where $\hat{h}_1$ has been defined in eqn 2.31 and the sum extends over all electrons apart from 1. Note that in order to obtain the v_{1j} all the other orbitals must be known in advance. This poses a circular problem since ϕ_1 must be known in order to calculate any of the other orbitals. The solution to this problem is to make an initial guess for all of the other orbitals and to use these 'zero-order' orbitals to calculate the expectation potential energy function for electron 1. The one-electron equation is then solved for ϕ_1, giving a first-order approximation. The new ϕ_1 is included in the calculation of the first-order approximation to ϕ_2 and all the electrons are cycled through. When this process is complete the first-order approximations are used to calculate a second-order approximation to ϕ_1, and the iteration is continued until the computed energy reaches a stable value. The number of iterations required depends on how good the initial guess was.

When the electron–electron repulsion is included in this way we have to be careful in calculating the total energy of the molecule. Each one-electron energy E_i includes a contribution from the electrostatic interaction with every other electron, including for example the repulsion ij. But the energy of electron j also includes the same electrostatic repulsion so that simply adding up the one-electron energies counts the repulsions twice over. The total expectation repulsion energy therefore has to be subtracted from the final summation:

$$E = \sum_i E_i - \frac{e^2}{4\pi\epsilon_o} \sum_{i<j} \int \frac{\phi_i^*\phi_i.\phi_j^*\phi_j}{r_{ij}} d\tau_i d\tau_j \tag{2.35}$$

Finally we point in the direction in which this subject develops. The product wavefunction written in eqn 2.32 has assigned electron 1 (coordinates $\mathbf{r}_1$) to orbital 1 (ϕ_1), electron 2 to orbital 2 and so forth. However, since the electrons are indistinguishable this assignment is arbitrary. It would be equally valid to permute electrons 1 and 2, putting electron 2 in orbital 1, and electron 1 in orbital 2, giving the wavefunction

$$\phi_1(\mathbf{r}_2)\phi_2(\mathbf{r}_1)\ldots\phi_n(\mathbf{r}_n)$$

Obviously every possible permutation of the electron labels leads to an equally valid solution with the same energy. Furthermore any linear combination of these permutations will also be a solution, but will have a different energy because of the way in which the electron repulsion energy has been patched into the approximation. For a solution to be physically meaningful no permutation of the electrons should change any physical property. In addition there is a fundamental symmetry rule, known as the Pauli principle, that the total electronic wavefunction (including spin) must change sign whenever two electrons are permuted. This leads to a more complicated formulation in which the electron wavefunction is represented by a linear combination of spin-orbital products with this property. The self-consistent field theory applied to such antisymmetric product functions is known as the *Hartree–Fock* theory, and will be discussed further in II.

2.8   The diatomic molecule

We now turn our attention to the separated Schrödinger eqn 2.27 for the nuclear wavefunction. The case of a diatomic molecule is considerably simpler than that of a polyatomic molecule, and so we discuss diatomic molecules first.

Suppose that the two nuclei in the diatomic molecule have masses m_1 and m_2 and their coordinates are $\mathbf{r}_1$ and $\mathbf{r}_2$. The nuclear Schrödinger equation will contain terms for the kinetic energies of both nuclei, and the potential energy $V(\mathbf{r}_1,\mathbf{r}_2)$, which includes the nucleus–nucleus repulsion energy and the averaged electronic energy discussed in Section 2.7.

$$-\frac{\hbar^2}{2m_1}\nabla_1^2\psi_n - \frac{\hbar^2}{2m_2}\nabla_2^2\psi_n + V(\mathbf{r}_1,\mathbf{r}_2)\psi_n = E\psi_n \qquad (2.36)$$

The diatomic molecule is effectively a problem involving two interacting particles, which suggests the approach discussed in Section 2.4, separating centre of mass coordinates and relative coordinates. This transformation will lead to a separation of variables if the potential energy only depends on the the interparticle vector, $\mathbf{r}_r = \mathbf{r}_2 - \mathbf{r}_1$, which will normally be the case in the absence of an external field, since the potential energy of a molecule does not then depend on the position of the molecule in space. We therefore find a transformed equation of the form

$$-\frac{\hbar^2}{2M}\nabla_c^2\psi_n - \frac{\hbar^2}{2\mu}\nabla_r^2\psi_n + V(\mathbf{r}_r)\psi_n = E\psi_n \qquad (2.37)$$

where M is the total mass, $m_1 + m_2$, and μ is the reduced mass, $m_1 m_2/M$. Separating the centre of mass coordinates from the relative coordinates we find a Schrödinger equation for the translational motion of the centre of mass and an equation analogous to eqn 2.23 for the internal nuclear wavefunction:

$$-\frac{\hbar^2}{2\mu}\nabla_r^2 R + V(\mathbf{r})R = E_{int}R \qquad (2.38)$$

The eigenvalue E_{int} represents the internal energy of the molecule, excluding the translational energy of the molecular centre of mass.

The second stage is to recognize that the potential energy of the molecule is also independent of its orientation in space, since space has the same properties in all directions. Thus the potential energy depends only on the distance between the two nuclei, r, and the system can be recognized as spherically symmetrical, thus permitting the simplification discussed in Section 2.3. The angular and radial variables can therefore be separated giving the solutions to the angular part of the wavefunction given in Chapter 3 and the radial Schrödinger equation from eqn 2.18

$$-\frac{\hbar^2}{2\mu r^2}\frac{\mathrm{d}}{\mathrm{d}r}\left(r^2\frac{\mathrm{d}R}{\mathrm{d}r}\right) + \left(V(r) + \frac{\hbar^2 J(J+1)}{2\mu r^2}\right)R = E_{int}R \qquad (2.39)$$

Motion in the angular variables θ and ϕ corresponds to rotation of the molecule, and this contributes the small rotational energy term $\hbar^2 J(J+1)/2\mu r^2$ to the potential energy for the radial equation. Motion in the radial coordinate describes variation of the distance between the two nuclei, i.e. vibration of the molecule. Notice that although the Schrödinger equation for the rotations and vibrations can be separated rigorously into equations in the angular variables and the radius, the energy of the molecule does not split up neatly into the sum of a rotational energy and a vibrational energy. This is because the angular momentum is a constant of the motion, and hence the rotational energy $\hbar^2 J(J+1)/2\mu r^2$ depends on the bond length r, and so has to be included in the potential energy part of the vibrational equation, although it is a relatively small correction.

Finally we make the substitution $S(r) = rR(r)$, as in eqn 2.19, and the vibrational equation becomes

$$-\frac{\hbar^2}{2\mu}\frac{\mathrm{d}^2 S}{\mathrm{d}r^2} + \left(V(r) + \frac{\hbar^2 J(J+1)}{2\mu r^2}\right)S = E_{int}S \qquad (2.40)$$

which is a one-dimensional Schrödinger equation with a potential energy $V(r) + \hbar^2 J(J+1)/2\mu r^2$. No further analytical progress can be made at this stage because the function $V(r)$ does not in general have any simple functional form, being the sum of the nucleus–nucleus repulsion energy and the electronic energy, as discussed in Section 2.7. Of course for a known potential energy curve $V(r)$ numerical solutions can be found by standard computational methods. For each value of the angular momentum quantum number J such a solution will give a stack of eigenvalues, labelled by a vibrational quantum number v and their corresponding eigenfunctions.

Problem 2.8.1. *Show how the normalization integrals of the radial function R and the transformed function S are related.*

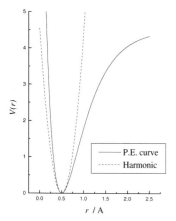

Fig. 2.3 Potential energy curve for the H_2 molecule compared with the harmonic approximation.

The harmonic oscillator-rigid rotor approximation

Although $V(r)$ does not have a simple functional form, very good representations of the experimentally observed energy levels can be obtained by taking approximate forms with sensible properties. Figure 2.3 represents the typical form of $V(r)$ for the ground state of a diatomic molecule. The minimum of the curve is the most stable configuration of the nuclei and corresponds to the equilibrium bond length r_e. At normal temperatures the molecule only has sufficient energy to sample the region around the bottom of the well, where the bond is only stretched by a small amount. Thus the molecule can be thought of as executing small amplitude vibrations about r_e. As a first approximation we can take a Taylor expansion of $V(r)$ about the point $r = r_e$.

$$V(r) = V(r_e) + V'(r_e)(r - r_e) + \frac{1}{2}V''(r_e)(r - r_e)^2 + \dots \quad (2.41)$$

$$= V_e + \frac{1}{2}k(r - r_e)^2 + \dots$$

The first derivative is equal to zero because r_e is a minimum of the potential energy curve, and the second derivative (or curvature) of the potential energy is a positive constant, denoted k and called the *force constant*. The use of a quadratic potential energy function is called the *harmonic approximation*. The quality of the fit is illustrated in Fig. 2.3 demonstrating that it is only really applicable for low energy states.

The rotational energy is also a function of r, but since the vibrations are of small amplitude and the rotational energy is only a small correction to the potential energy curve, it is usual to approximate the rotational energy to $\hbar^2 J(J+1)/2\mu r_e^2$, neglecting its dependence on the bond length. In this *rigid rotor* approximation the internal energy of the molecule separates into the sum of the rigid rotor rotational energy and the vibrational energy.

2.9 Vibrations of a polyatomic molecule

[This section is more advanced.] To specify the exact positions of both nuclei in a diatomic molecule it is necessary to specify three coordinates for each nucleus. The number of independent variables required, or *degrees of freedom* cannot be altered by any coordinate transformation. Three degrees of freedom can be assigned to specify the position of the centre of mass of the molecule, and two more (θ and ϕ) are needed to specify the orientation of the molecule, leaving one vibrational degree of freedom (r) to specify the internal configuration of the molecule. The case of a nonlinear polyatomic molecule of N atoms is similar—of the $3N$ coordinates needed to specify the exact positions of all N nuclei, three degrees of freedom are taken up by specifying the centre of mass, but three angles are required to specify the orientation, leaving $3N - 6$ vibrational degrees of freedom. In this section we consider these vibrational degrees of freedom. Since polyatomic molecules are significantly more complex than diatomic molecules we restrict ourselves to the rigid rotor harmonic oscillator approximation so that we can consider the rotational energy and the vibrational energy separately in the first instance. Here we concentrate solely on vibrations; rotational energies are dealt with in Chapter 3.

The Schrödinger equation for the vibrational wavefunction relies on a knowledge of the potential energy V as a function of the nuclear coordinates. The equilibrium nuclear configuration will be found at a minimum of this potential energy function and the vibrations will be treated as small amplitude perturbations of this configuration. It is the existence of such a minimum that permits us to talk about the shape or geometry of a molecule.

The potential energy can be expanded in a Taylor series about the equilibrium configuration. Denoting the deviation of coordinate i from equilibrium by x_i the potential energy function becomes

$$V = V_o + \sum_{i,j} \frac{1}{2} \frac{\partial^2 V}{\partial x_i \partial x_j} x_i x_j + \ldots = V_o + \sum_{i,j} \frac{1}{2} k_{ij} x_i x_j + \ldots \quad (2.42)$$

Notice all the first derivatives of V are zero because the equilibrium configuration is the minimum potential energy, and that $k_{ij} = k_{ji}$.

Before considering the Schrödinger equation for this system we consider the classical equations of motion for the same potential energy. Application of Newton's second law gives

$$m_i \ddot{x}_i = -\sum_j k_{ij} x_j \quad (2.43)$$

This is a set of coupled differential equations, each double derivative being weighted by the mass of its respective nucleus. Firstly we get rid of the mass weightings by transforming to mass-weighted coordinates $q_i = \sqrt{m_i} x_i$, which brings the equation into the form

$$\ddot{q}_i = -\sum_j \frac{k_{ij}}{\sqrt{m_i m_j}} q_j = -\sum_j F_{ij} q_j \quad (2.44)$$

The matrix $\mathbf{F}$, whose elements F_{ij} are defined by this equation, is symmetric, and can therefore be transfomed to a diagonal matrix $\mathbf{\Lambda}$ by an orthogonal transformation $\mathbf{U}$, so that $\mathbf{U}^T \mathbf{F} \mathbf{U} = \mathbf{\Lambda}$. Thus

$$\ddot{q}_i = -\sum_j (U \Lambda U^T)_{ij} q_j \quad (2.45)$$

or introducing the coordinates defined by $Q_i = \sum_j U_{ji} q_j$

$$\ddot{Q}_i = -\sum_j \Lambda_{ij} Q_j = -\Lambda_i Q_i \quad (2.46)$$

This is the classical equation of motion for a simple harmonic oscillator with angular frequency $\sqrt{\Lambda_i}$. We can see that by first mass-weighting the coordinates and then diagonalizing the matrix $\mathbf{F}$ the classical problem is reduced to a set of uncoupled harmonic oscillators. The coordinates which effect this uncoupling are called *normal coordinates*.

The same is true of the quantum mechanical problem. The nuclear Schrödinger equation is of the form

$$-\sum_i \frac{\hbar^2}{2m_i}\nabla_i^2\psi_n + \sum_{i,j}\frac{1}{2}k_{ij}x_ix_j\psi_n = E_n\psi_n \tag{2.47}$$

after mass weighting this becomes

$$-\sum_i \frac{\hbar^2}{2}\nabla_i^2\psi_n + \sum_{i,j}\frac{1}{2}F_{ij}q_iq_j\psi_n = E_n\psi_n \tag{2.48}$$

removing the mass from the kinetic energy terms, and following diagonalization we get

$$-\sum_i \frac{\hbar^2}{2}\nabla_i^2\psi_n + \sum_i\frac{1}{2}\Lambda_iQ_i^2\psi_n = E_n\psi_n \tag{2.49}$$

where the differentiation is now with respect to the normal coordinates Q_i. Separation of variables now gives a single equation in each of the normal coordinates. The vibrational wavefunction can therefore be factorized in the form

$$\psi_n(Q_1, Q_2, \ldots) = \prod_i \phi_i(Q_i) \tag{2.50}$$

with a separated equation for each of the constituent oscillators:

$$-\frac{\hbar^2}{2}\nabla_i^2\phi_i + \frac{1}{2}\Lambda_iQ_i^2\phi_i = E_i\phi_i \tag{2.51}$$

which is a Schrödinger equation for a simple harmonic oscillator of frequency $\sqrt{\Lambda_i}$ (see Section 3.3). We conclude therefore that the vibrational wavefunction of a polyatomic molecule can be decomposed into a product of wavefunctions for simple harmonic oscillators, one in each mode. This simplification is of great importance in understanding the vibrational spectra of molecules.

Problem 2.9.1. *Consider the motions of the CO_2 molecule in the z-direction. These should comprise motion of the centre of mass in the z-direction, the symmetric stretch and the asymmetric stretch. Assuming the two bonds to be harmonic with a force constant k find expressions for the two vibrational frequencies and their associated motions.*

References

1. Stephenson, G. (1973). *Mathematical methods for science students.* Section 24.4, p. 469. Longman, London.
2. Levine, R. B. and Bernstein, R. B. (1987). *Molecular reaction dynamics and chemical reactivity.* Oxford University Press, Oxford.
3. Bunker, P. R. (1979). *Molecular symmetry and spectroscopy.* Academic Press, San Diego.

3 Exact solutions

In Chapter 1 we showed how a Schrödinger equation can be set up for atoms and molecules by using the coordinates of the nuclei and electrons as variables. It is difficult to gain any physical insight from the solution to such a complicated problem, and so Chapter 2 detailed several ways in which this description can be simplified by separating the variables upon which the wavefunction depends. These separations transform the problem from a single multi-dimensional equation to a set of one-variable equations. In some cases the simplification is a fundamental, exact property of the system; in other cases (for example, the Born–Oppenheimer separation, or the orbital approximation) it is approximate.

In this chapter we consider some of the simplified Schrödinger equations obtained in Chapter 2, and find explicit solutions for the energy levels and the corresponding wavefunctions.

3.1   The free particle

In Section 2.4 we separated the coordinates of the centre of mass of an atom or a molecule, leading to a Schrödinger equation for the translational motion:

$$-\frac{\hbar^2}{2M}\nabla^2\psi = E\psi \tag{3.1}$$

This equation can be separated further, since the Cartesian variables x, y, z can be separated, and a solution found of the form $X(x)Y(y)Z(z)$. The separated equation for X is

$$-\frac{\hbar^2}{2M}\frac{\mathrm{d}^2 X}{\mathrm{d}x^2} = E_x X(x) \tag{3.2}$$

with similar equations for Y and Z. These are simple second-order differential equations with constant coefficients, whose solutions are well known to be

$$X(x) = A\exp(\mathrm{i}p_x x/\hbar) \tag{3.3}$$

where $p_x = \sqrt{2ME_x}$ is the x-component of the momentum, and A is an arbitrary constant. These functions are also eigenfunctions of the momentum operator, as discussed in Section 1.3.

Combining all three parts of the solution, we obtain the wavefunction

$$\psi = N\exp(\mathrm{i}(p_x x + p_y y + p_z z)/\hbar) = N\exp(\mathrm{i}\mathbf{p}.\mathbf{r}/\hbar) \tag{3.4}$$

where $\mathbf{p}$ represents the momentum vector and $\mathbf{r}$ the position vector of the particle. This equation is the three-dimensional momentum eigenfunction, as claimed in Section 1.2. N is a normalization constant.

3.2 Particle in a box

Problem 3.2.1. *Show that the only solutions of eqn 3.2 that obey both boundary conditions are*
$\psi_n = N\sin(n\pi x/a)$, *with*
$n = 1, 2, 3, \ldots$, *and that the energy is*
$E_n = n^2 h^2/8Ma^2$.

The *particle in a box* of Problem 1.3.1 is a related problem. The wavefunction is also a solution to eqn 3.2. The only difference is that a free particle is not confined by a boundary, whereas the particle in a box must be found inside the box between $x = 0$ and $x = a$. This means that ψ must be zero everywhere outside the box, and since ψ is continuous, it must also be zero at the edges of the box.

The solutions can also be written in the form

$$\psi = \frac{N}{2i}\left[e^{in\pi x/a} - e^{-in\pi x/a}\right] \tag{3.5}$$

so that inside the box the wavefunction is a superposition of two free-particle wavefunctions with characteristic momenta $\pm nh/2a$. This observation means essentially that the absolute value of the momentum is certain, but that the particle is equally likely to be travelling to the right or to the left. A more rigorous analysis of the momentum in this case, including boundary effects, can be found in reference 1.

The particle in a box is the simplest example of a confined particle, and enables us to understand the origin of many features of such systems, such as the appearance of discrete energy levels. These arise from the restriction of n to be a whole number, which is forced by the boundary conditions describing the confinement of the particle, insisting that the wavefunction be zero at both edges of the box. It is a general feature of particles confined by a potential to some region of space that solutions to the Schrödinger equation can only be found at certain discrete energies.

Another feature of bound systems is the existence of a *zero-point energy*, which is the lowest energy that the particle can have. The origin of the zero-point energy can be understood in terms of the uncertainty principle. Since the particle is confined in some region of space, its position has a finite uncertainty. This means in turn that its momentum must have a nonzero uncertainty. If a particle could exist in a restricted region of space with zero kinetic energy, then its momentum would be exactly zero, and the uncertainty principle would be violated. Thus, a confined particle must have nonzero kinetic energy.

3.3 The harmonic oscillator

We now turn to another type of system confined by a potential, the harmonic oscillator, which was introduced in Section 2.8 as an approximate model for the vibration of a diatomic molecule. The harmonic oscillator shows similar features to the particle in a box, but the detailed solution is more difficult. It is not necessary to follow all the details of the mathematical development which follows, but all chemistry students should be familiar with the forms of the resulting wavefunctions and the energy levels.

The harmonic oscillator Schrödinger equation is

$$-\frac{\hbar^2}{2\mu}\frac{\partial^2\psi}{\partial x^2} + \frac{1}{2}kx^2\psi = E_v\psi(x) \tag{3.6}$$

In the application to the vibration of a diatomic molecule $x = r - r_e$, the bond extension from its equilibrium value r_e. For polyatomic molecules x represents a normal vibrational coordinate as described in Section 2.9.

In classical mechanics a particle in a parabolic potential well, $\frac{1}{2}kx^2$, experiences a restoring force $-kx$ towards the point 0, and undergoes a sinusoidal motion called *simple harmonic motion* with angular frequency $\omega = \sqrt{k/\mu}$.

Ladder operator method

(*This subsection is more advanced; for an alternative, see reference 2.*) Following the method of Section 2.9 we introduce mass-weighted coordinates $q = x\sqrt{\mu}$. The Schrödinger equation then becomes

$$\frac{1}{2}(\hat{p}^2 + \omega^2 \hat{q}^2)\psi = E\psi \tag{3.7}$$

where $\hat{q}$ is the position operator and $\hat{p}$ is the momentum operator, $\hat{p} = -i\hbar d/dq$. We now define the operators $\hat{R}^+$ and $\hat{R}^-$,

$$\hat{R}^{\pm} = \frac{1}{\sqrt{2}}[\hat{p} \pm i\omega\hat{q}] \tag{3.8}$$

The commutator of Problem 3.3.1 can be rearranged to find

$$\hat{H}(\hat{R}^{\pm}\psi) = (E \pm \hbar\omega)(\hat{R}^{\pm}\psi) \tag{3.10}$$

Thus, if ψ is an eigenfunction of $\hat{H}$ with energy E, then $\hat{R}^{\pm}\psi$ are also eigenfunctions of $\hat{H}$ with energies shifted by $\pm\hbar\omega$. $\hat{R}^+$ therefore transforms a wavefunction with energy E to a wavefunction with energy $E + \hbar\omega$ and is called a *ladder-up* operator. $\hat{R}^-$ ladders the energy down by the same constant amount $\hbar\omega$.

The results of Problems 3.3.2 and 3.3.3 can be combined: if ψ is an eigenfunction of $\hat{H}$, then

$$E = \langle \psi|\hat{H}|\psi \rangle = \langle \psi|\hat{R}^+\hat{R}^-|\psi \rangle + \frac{1}{2}\hbar\omega \tag{3.11}$$

but because the expectation of $\hat{R}^+\hat{R}^-$ cannot be negative,

$$E \geq \frac{1}{2}\hbar\omega \tag{3.12}$$

Problem 3.3.1. *Use the results of Problem 1.4.1 to show that the commutators $[\hat{H}, \hat{R}^{\pm}]$ are given by*

$$[\hat{H}, \hat{R}^{\pm}] = \pm\hbar\omega\hat{R}^{\pm} \tag{3.9}$$

Problem 3.3.2. *Show that $\hat{H} = \hat{R}^+\hat{R}^- + \frac{1}{2}\hbar\omega$.*

Problem 3.3.3. *Show that, since $\hat{R}^+$ and $\hat{R}^-$ are hermitian conjugates, the expectation $\langle \psi|\hat{R}^+\hat{R}^-|\psi \rangle$ must be real and non-negative.*

Applying the ladder-down operator reduces the energy by $\hbar\omega$. But, since no permissible energy can be less than $\frac{1}{2}\hbar\omega$, it cannot be possible to carry on laddering down for ever. There must come a point when laddering down would take E below $\frac{1}{2}\hbar\omega$, which is not possible. We conclude that there must be a lowest possible value of the energy, E_0, with a corresponding eigenfunction ψ_0, and that since it is not possible to ladder down from this energy $\hat{R}^-\psi_0 = 0$.

Applying the explicit form of $\hat{R}^-$, we obtain a first-order differential equation for the lowest wavefunction, ψ_0

$$\hbar\frac{\mathrm{d}\psi_0}{\mathrm{d}q} + \omega q\psi_0 = 0 \tag{3.13}$$

which can be integrated to give

$$\psi_0(q) = N\exp(-\omega q^2/2\hbar) \tag{3.14}$$

The energy of this state, which is the lowest permissible energy, can be found by application of the Hamiltonian

$$\hat{H}\psi_0 = \hat{R}^+\hat{R}^-\psi_0 + \frac{1}{2}\hbar\omega\psi_0 \tag{3.15}$$

The first term is zero because $\hat{R}^-\psi_0 = 0$, and therefore we conclude that $E_0 = \frac{1}{2}\hbar\omega$.

Higher energy levels can be found by laddering up, so that the vth energy level above E_0 has energy

$$E_v = (v + \frac{1}{2})\hbar\omega \tag{3.16}$$

and the corresponding wavefunction is

$$\psi_v = (\hat{R}^+)^v\psi_0 \tag{3.17}$$

Of course ψ_v obtained in this way still has to be normalized.

Problem 3.3.4. *Verify that application of $\hat{R}^+$ to ψ_0 followed by normalization gives ψ_1 as tabulated in Table 3.1. [You will need the following standard integral to perform the normalization]*

$$\int_{-\infty}^{\infty} x^{2n}\exp(-\alpha^2 x^2) = \frac{(2n)!\sqrt{\pi}}{n!(2\alpha)^{2n+1}}$$

No other solutions with energies between $\hbar\omega/2$ and $3\hbar\omega/2$ are possible since ψ_0 is the only solution of the equation $\hat{R}^-\psi_0 = 0$.

Properties of the solutions

The wavefunctions for the harmonic oscillator have several interesting features.

1. Like the particle in a box, the harmonic oscillator is a bound system, and so solutions only exist at certain well-defined energy levels. The energy levels form a ladder with an equal spacing of $\hbar\omega$, which is h times the vibrational frequency of a classical harmonic oscillator with the same potential energy function (see Fig. 3.1).

2. There is a *zero-point energy* equal to $\frac{1}{2}\hbar\omega$. Zero-point energy has important consequences in chemistry: for example it is the main source of the difference between the dissociation energies of the H_2 and the D_2 molecules. Both molecules have the same potential energy curves, and therefore the same force constants, because the electronic wavefunctions which produce the bonding are not dependent on the nuclear masses. However, the vibrational frequencies

Table 3.1 Harmonic oscillator wavefunctions

v	ψ_v
0	$N_0\exp(-\xi^2/2)$
1	$N_1 2\xi\exp(-\xi^2/2)$
2	$N_2(4\xi^2 - 2)\exp(-\xi^2/2)$
3	$N_3 4\xi(2\xi^2 - 3)\exp(-\xi^2/2)$

$$N_v = (2^n n!)^{-\frac{1}{2}}(\mu\omega/\pi\hbar)^{\frac{1}{4}}$$
$$\xi = q(\omega/\hbar)^{\frac{1}{2}}$$

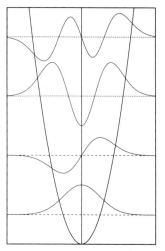

Fig. 3.1 Vibrational wavefunctions and energy levels for the harmonic oscillator.

Zero-point energy is a illustration of the uncertainty principle. Because the particle is confined in the potential well there is a finite uncertainty in its position.

(and therefore the zero point energies) are different because they do depend on the nuclear masses ($\omega = \sqrt{k/\mu}$); in consequence, the bond dissociation energy of D_2 is larger than that of H_2, as shown in Fig. 3.2.

3. The wavefunctions are depicted in Fig. 3.1. The lowest energy wavefunction contains no nodes, and is an even function. ψ_1 has one node and is an odd function of q, etc. In each case the wavefunction penetrates into classically forbidden regions where the total energy is less than the potential energy. There is therefore a nonzero probability of finding the oscillator in these classically forbidden regions. This *tunnelling* has important consequences in chemistry, leading to the possibility of penetrating barriers even when there does not seem to be enough energy available.

4. The probability density is depicted in Fig. 3.3. In the lowest energy level the wavefunction is concentrated about the centre of the well at $q = 0$, which represents the equilibrium bond extension. As we climb higher up the ladder the wavefunction has higher amplitude further and further away from 0. At high energy the probability density gradually approaches the classical limit where the oscillator is most likely to be found near its turning points, since it spends most time in regions of space where it is travelling slowest.

The Morse potential

The harmonic approximation is not exact, as can be seen by reference to Fig. 2.3. The parabolic potential function may be a good approximation close to the minimum of the curve, but it exhibits two unphysical tendencies further away from the minimum.

1. The true potential energy becomes infinite as the bond length approaches zero, whereas the harmonic approximation is finite and the wavefunctions are nonzero in the impossible region $r < 0$.

2. At long bond extensions the potential energy in a real molecule approaches an asymptote, corresponding to two dissociated atoms, whereas the harmonic approximation shows no such effect.

The broadening of the potential well towards dissociation leads to a reduction in the the energy level spacing as the energy increases (see Fig. 3.4). The energy levels finally converge into a continuum at the dissociation limit. For such systems an improved approximation to the potential energy function is the Morse potential

$$V(r) = D_e[1 - \exp(-\beta(r - r_e))]^2 \tag{3.18}$$

This function answers the second criticism, tending to a constant dissociation energy D_e at high bond extensions. However, the potential function is still finite (but large) at $r = 0$. For the Morse potential the vibrational energy levels become

There is therefore a minimum uncertainty in the momentum, which means that the particle must have a positive expectation kinetic energy, and therefore a positive energy.

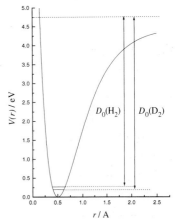

Fig. 3.2 Potential energy curve for the H_2 molecule, showing the different zero-point energies and dissociation energies of H_2 and D_2.

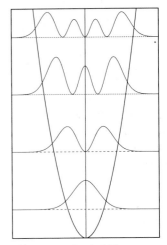

Fig. 3.3 Probability densities ψ^2 for the harmonic oscillator.

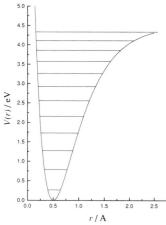

Fig. 3.4 Morse potential and its energy levels for the hydrogen molecule.

$$E_v = \hbar\omega_e(v + \frac{1}{2}) - \hbar\omega_e x_e(v + \frac{1}{2})^2 \qquad (3.19)$$

This energy level formula is often used to analyse vibrational structure in infra-red and electronic spectroscopy.[3] The parameter x_e is known as the anharmonicity.

3.4 Angular momentum

Angular momentum is an extremely important concept in chemistry. It is not possible to understand most types of spectroscopy, for example, without a grasp of the quantum mechanical nature of angular momentum, and in particular its eigenvalues and eigenfunctions.

Spherical symmetry

In a spherically symmetrical system the potential energy does not depend on either of the spherical polar angle variables θ or ϕ, and in classical mechanics the angular momentum *vector* is a constant of the motion. We have already considered the angular momentum operator for such a system (eqn 1.20), and the separation of its eigenfunctions into products of the form $Y(\theta, \phi) = \Theta(\theta)\Phi(\phi)$ (Section 2.3). We now consider the properties of the angular momentum in more detail.

Angular momentum operators

First we formulate operators for the three components of the angular momentum, $\hat{l}_x$, $\hat{l}_y$ and $\hat{l}_z$. The operators are calculated from the correspondence principle using eqn 1.18. For example, we have

$$\hat{l}_z = \hat{x}\hat{p}_y - \hat{y}\hat{p}_x \qquad (3.20)$$

In the context of the uncertainty principle the commutators of Problems 3.4.1 and 3.4.2 imply that it is possible to specify simultaneous values for the squared magnitude of the angular momentum and any *one* component. But, because the three component operators do not commute with one another it is not possible to specify more than one component at the same time, except for the special case where the angular momentum is precisely zero. The maximum information about the angular momentum permitted is therefore the eigenvalue of the squared magnitude and one component, which we arbitrarily choose to be the z-component.

Problem 3.4.1. *Applying the rules for commutators derived in Problem 1.4.1, demonstrate the following:*

$$[\hat{l}_x, \hat{l}_y] = i\hbar\hat{l}_z,$$
$$[\hat{l}_y, \hat{l}_z] = i\hbar\hat{l}_x, \quad (3.21)$$
$$[\hat{l}_z, \hat{l}_x] = i\hbar\hat{l}_y$$

[Hint: it is only necessary to derive one of these, the others follow by cyclic permutation of x, y and z.]

Problem 3.4.2. *Defining the operator for the square of the total angular momentum by $\hat{l}^2 = \hat{l}_x^2 + \hat{l}_y^2 + \hat{l}_z^2$, show that*

$$[\hat{l}_x, \hat{l}^2] = [\hat{l}_y, \hat{l}^2] = [\hat{l}_z, \hat{l}^2] = 0 \qquad (3.22)$$

[The algebra is much simplified if the results of Problem 1.4.1 are used.]

Separation of variables

We already know the eigenfunctions and eigenvalues of the operator $\hat{l}_z$. The operator is most conveniently represented in polar coordinates:

$$\hat{l}_z = -i\hbar\frac{\partial}{\partial\phi} \qquad (3.23)$$

which we recognize as the angular momentum operator pertaining in cylindrical symmetry, whose eigenfunctions are $\exp(\pm im_l\phi)$, with corresponding eigenvalues $\pm\hbar m_l$.

In Chapter 2 we discussed the separation of the squared angular momentum operator in spherical polar coordinates. The separated equation in the variable ϕ is eqn 2.16

$$\frac{d^2\Phi}{d\phi^2} = -m_l^2\Phi \tag{3.24}$$

This is the eigenvalue equation for the square of the $\hat{l}_z$ operator, which has the same eigenfunctions as $\hat{l}_z$.

The equation in θ (2.18) is harder to solve. As for the harmonic oscillator there are two possible methods. For a detailed solution by the method of series see reference 4. The solutions have been known since the early nineteenth century, and are known as associated Legendre polynomials. The mathematical details are not important, but the form of the solutions and the possible values for the separation constant W (see eqn 2.14) are very important. Solutions only exist if $W = \hbar^2 l(l + 1)/2m$ where l is an integer $\geq |m_l|$. In other words, for a given value of l, m_l may take whole number values only from $-l$ to $+l$. This means that in a spherical system the square of the angular momentum may only take values $\hbar^2 l(l+1)$, and that this value has a degeneracy of $2l+1$ corresponding to the $2l + 1$ values of m_l between $-l$ and l. The first few eigenfunctions will be found in Table 3.2.

When the functions Θ and Φ are multiplied together, the resulting solutions are known as *spherical harmonics*. They are the fundamental angular solutions of the Schrödinger equation for all problems which are equivalent to a single entity moving in a spherically symmetrical potential. They are also of great utility in other areas of science, for example in geology they are used to analyse the vibration of the earth. Some of them are depicted in Fig. 3.5. They should appear familiar; they also control the shapes of the atomic orbitals because atoms are effectively spherically symmetrical.

One of the most important applications of the spherical harmonics in chemistry is to the rotation of diatomic molecules. Rotation of a rigid molecule simply involves motion of its orientation, specified by the angles θ and ϕ. In an isotropic space the potential energy of a molecule is independent of its orientation, and so the rotational wavefunctions are spherical harmonics, and the square of the rotational angular momentum takes the values $\hbar^2 J(J + 1)$ (J is used instead of l to signify the total angular momentum quantum number in this context). The classical rotational kinetic energy is equal to $l^2/2I$, where I is the moment of inertia: using the correspondence principle the corresponding rotational Hamiltonian is $\hat{l}^2/2I$, and the rotational energy levels follow immediately:

$$E_J = \frac{\hbar^2}{2I}J(J+1) \tag{3.25}$$

each of which has a degeneracy of $2J + 1$.

Table 3.2 Rotational wavefunctions, $\Theta_{lm}(\theta)$

l	m	$\Theta_{lm}(\theta)$
0	0	$\left(\frac{1}{2}\right)^{1/2}$
1	0	$\left(\frac{3}{2}\right)^{1/2}\cos\theta$
1	± 1	$\left(\frac{3}{4}\right)^{1/2}\sin\theta$
2	0	$\left(\frac{5}{8}\right)^{1/2}(3\cos^2\theta - 1)$
2	± 1	$\left(\frac{15}{4}\right)^{1/2}\sin\theta\cos\theta$
2	± 2	$\left(\frac{15}{16}\right)^{1/2}\sin^2\theta$

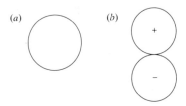

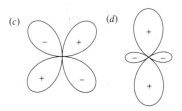

Fig. 3.5 Polar diagrams for some spherical harmonics: (a) $l = 0$, s orbital; (b) $l = 1$, $m_l = 0$, p_z orbital; (c) $l = 2$, $m_l = 1$, d orbital; (d) $l = 2$, $m_l = 0$, d_{z^2} orbital.

Ladder operator method

[This subsection is more advanced.]

Just as for the harmonic oscillator, there is an elegant method for solution of the θ equation, using ladder operators. The required operators, $\hat{l}_\pm$, are defined by

$$\hat{l}_\pm = \hat{l}_x \pm i\hat{l}_y \tag{3.26}$$

Problem 3.4.3. *Show using eqns 3.21 and 3.22 that the commutators of $\hat{l}_\pm$ with $\hat{l}^2$ and $\hat{l}_z$ are given by $[\hat{l}^2, \hat{l}_\pm] = 0$ and $[\hat{l}_z, \hat{l}_\pm] = \pm\hbar l_\pm$.*

The first of the commutators of Problem 3.4.3 implies that the operation of $\hat{l}_\pm$ on an eigenfunction of $\hat{l}^2$ does not alter the eigenvalue. Applying the second commutation relation for $\hat{l}_+$ to an eigenfunction of l_z, which we label $|M>$, we find

$$\hat{l}_z\hat{l}_+|M> = \hat{l}_+\hat{l}_z|M> + \hbar\hat{l}_+|M> = \hbar(M+1)\hat{l}_+|M> \tag{3.27}$$

In other words $\hat{l}_+|M>$ is an eigenfunction of $\hat{l}_z$ whose eigenvalue has been laddered up from $\hbar M$ to $\hbar(M+1)$, (cf. $\hat{R}^+$ in Section 3.3).

Problem 3.4.4. *Show that $\hat{l}_-$ ladders the eigenvalue down by $\hbar$.*

The ladder operators can also be represented in spherical polar coordinates:

$$\hat{l}_\pm = \hbar\exp(\pm i\phi)\left(\pm\frac{\partial}{\partial\theta} + i\cot\theta\frac{\partial}{\partial\phi}\right) \tag{3.28}$$

The eigenvalue of $\hat{l}_z^2$ may not be larger than that of $\hat{l}^2$, because the component of a vector in a particular direction may not be larger than the length of the whole vector. In consequence, for a given eigenvalue of $\hat{l}^2$ it cannot be possible to continue laddering up indefinitely; there must be a maximum possible value of l_z. Let the maximum possible value of M for the given total angular momentum be L. The application of $\hat{l}_+$ on the state $|L>$ must give zero, since it is not possible to ladder up any further. We now use this fact to derive the eigenvalues and eigenfunctions of $\hat{l}^2$.

Problem 3.4.5. *Show that $\hat{l}_-\hat{l}_+ = \hat{l}^2 - \hat{l}_z^2 - \hbar\hat{l}_z$*

Since $\hat{l}_+|L> = 0$, we can see immediately that $\hat{l}_-\hat{l}_+|L> = 0$, and hence (from Problem 3.4.5) $(\hat{l}^2 - \hat{l}_z^2 - \hbar\hat{l}_z)|L> = 0$, so that

$$\hat{l}^2|L> = (\hat{l}_z^2 + \hbar\hat{l}_z)|L> = \hbar^2 L(L+1)|L> \tag{3.29}$$

Problem 3.4.6. *Show that the minimum value of the z-component for the total angular momentum $\hbar L(L+1)$ is $-\hbar L$.*

We conclude that if the maximum z-component of the angular momentum is $\hbar L$, then the squared total angular momentum is $\hbar^2 L(L+1)$.

The angular momentum eigenfunctions can therefore be labelled with two integers: M specifies the z-component and L specifies the squared amplitude and hence the maximum value allowed for $|M|$. The eigenfunction is usually specified $|L, M>$.

The ladder operators may also be used to find the angular momentum eigenfunctions explicitly. Since

$$\hat{l}_+|L, L> = 0 \tag{3.30}$$

using the representation of $\hat{l}_+$ in polar coordinates, eqn 3.28, we find, after separating the variables θ and ϕ,

$$\hat{l}_+ |L, L> = \hbar \exp(i\phi) \left(\frac{\partial}{\partial \theta} + i \cot \theta \frac{\partial}{\partial \phi} \right) \Theta_{L,L}(\theta) \Phi_L(\phi) = 0 \qquad (3.31)$$

Now we substitute in the known form of $\Phi_L(\phi) = \exp(iL\phi)$ to obtain a first-order differential equation for Θ

$$\frac{d\Theta_{L,L}}{d\theta} = L \cot \theta \; \Theta_{L,L} \qquad (3.32)$$

which has the solution (subject to normalization)

$$\Theta_{L,L}(\theta) = N \sin^L \theta \qquad (3.33)$$

Exactly the same solution is found for $\Theta_{L,-L}$. Solutions for intermediate values of M can be found directly by using the ladder operators.

Problem 3.4.7. *Show that the tabulated eigenfunction $|2, 0 >$ can be found, both by laddering down twice from $|2, 2 >$, and by laddering up twice from $|2, -2 >$.*

3.5 Rotations of a polyatomic molecule

Rotational energy

A nonlinear molecule has three moments of inertia, I_a, I_b and I_c which describe the rotation of the molecule about its three orthogonal principal axes. If the components of the classical angular momentum about each axis are J_a, J_b and J_c, respectively, then the classical rotational energy of the molecule is

$$E_r = \frac{J_a^2}{2I_a} + \frac{J_b^2}{2I_b} + \frac{J_c^2}{2I_c} \qquad (3.34)$$

In the Schrödinger equation for the rotational wavefunction these classical angular momenta are replaced by the corresponding operators. However, the resulting equation is only easy to solve in special cases, because, although we can find eigenvalues for the total angular momentum and its component in one direction, we are precluded from knowing its simultaneous values in either of the other directions by the uncertainty principle (see Problems 3.4.1 and 3.4.2). We therefore only consider special cases here.

Symmetric top

In the case of the *symmetric top*, two of the principal moments of inertia are identical, for example $I_b = I_a$, so that the classical rotational energy can be written in terms of the total angular momentum J and one component, J_c,

$$E_r = \frac{J^2}{2I_a} + \left(\frac{1}{2I_c} - \frac{1}{2I_a} \right) J_c^2 \qquad (3.35)$$

The energy levels can therefore be written in terms of the eigenvalues of the operator for the total angular momentum squared, $\hbar^2 J(J+1)$, and the projection onto the c-axis in the molecule, $\hbar K$:

$$E_r = \frac{\hbar^2 J(J+1)}{2I_a} + \left(\frac{1}{2I_c} - \frac{1}{2I_a}\right)\hbar^2 K^2 \qquad (3.36)$$

In addition to the obvious degeneracy between positive and negative values of K in this equation, there is an additional degeneracy because, although the projection K has been specified along the c-axis *fixed in the molecule*, there is no link between this and an axis *fixed in space*. Thus each energy level with specified J and K has an additional space degeneracy of $2J+1$ corresponding to the different possible projections of J onto the space-fixed z-axis.

Spherical top

The second special case is known as a *spherical top*, in which all three moments of inertia are the same. In this case the rotational energy is simply

$$E_r = \frac{\hbar^2 J(J+1)}{2I_a} \qquad (3.37)$$

because $I_c = I_a$ in eqn 3.36. Each energy level is $(2J+1)^2$ degenerate.

Asymmetric top

The final possibility is that all three moments of inertia are different, and is beyond the scope of this book. The interested reader will find a detailed discussion in reference 5.

3.6 The hydrogen atom

Electronic wavefunctions for the H atom, or orbitals, can be written in the form $R(r)\Theta(\theta)\Phi(\phi)$.

To conclude this chapter we consider briefly the exact solutions to the Schrödinger equation for the hydrogen atom, in which the potential energy is spherically symmetrical. In Chapter 2 we have already considered the separation of the centre of mass coordinates, and the separation of the angular coordinates from the radial coordinates. In Section 3.5 we have just dealt with the solution to the angular part of the Schrödinger equation, leaving, as the final stage, the radial part of the equation:

$$-\frac{\hbar^2}{2\mu}\left(\frac{\partial^2 R}{\partial r^2} + \frac{2}{r}\frac{\partial R}{\partial r}\right) - \frac{e^2}{4\pi\epsilon_0 r}R + \frac{\hbar^2 l(l+1)}{2\mu r^2}R = ER \qquad (3.38)$$

where $\hbar^2 l(l+1)$ is the orbital angular momentum of the electron. Once again, the eigenvalues of this equation can be found by two methods, a classical method involving solution of the radial equation in power series and an operator method. The details are beyond the scope of this book, but the solutions are tabulated in Table 3.3. The interested reader may verify by direct substitution that the tabulated functions are indeed solutions of the radial equation. The solutions are labelled by a new quantum number n, called the *principal quantum number*. Solutions can only be found for values of n greater than l; equivalently for a given

Table 3.3 Hydrogen-like radial wavefunctions, $R_{nl}(r)$

n	l	$R_{nl}(r)$
1	0	$N \exp(-\rho)$
2	0	$N \left(\frac{1}{4\sqrt{2}} \right) (2 - \rho) \exp(-\rho/2)$
2	1	$N \left(\frac{1}{4\sqrt{2}} \right) \rho \exp(-\rho/2)$
3	0	$N \left(\frac{1}{81\sqrt{3}} \right) (27 - 18\rho + 2\rho^2) \exp(-\rho/3)$
3	1	$N \left(\frac{\sqrt{2}}{81} \right) \rho(6 - \rho) \exp(-\rho/3)$
3	2	$N \left(\frac{1}{81\sqrt{6}} \right) \rho^2 \exp(-\rho/3)$

Z is the nuclear charge, a is the Bohr radius,
$N = (Z^3/a^3)^{\frac{1}{2}}$, $\rho = Zr/a$

value of n the angular momentum quantum number l can only take values from 0 to $n - 1$.

The energy levels of the hydrogen atom turn out to depend only on the principal quantum number, and not on the angular momentum quantum number l.

$$E_n = -\frac{\mu e^4}{2(4\pi\epsilon_0)^2 n^2} \tag{3.39}$$

The reason for this additional degeneracy is not obvious—it is unique to potential energy functions which are proportional to r^{-1}, such as the Coulomb potential. Each energy level is labelled with a value for n, and has a degeneracy arising from the n possible values of l. In turn each of these values of l have a degeneracy $2l + 1$ arising from the $2l + 1$ possible values of the z-component of the angular momentum, m_l, between $-l$ and $+l$. It is straightforward to show that the overall degeneracy of the energy level n is equal to n^2:

$$g_n = \sum_{l=0}^{n-1} (2l + 1) = n^2 \tag{3.40}$$

(the sum is an arithmetic progression).

The radial and angular solutions tabulated in Tables 3.2 and Table 3.3 are multiplied together to give the full electronic wavefunctions of the H atom, or orbitals. For example the 3p orbital with $m_l = 1$ is given by

$$\psi = \frac{\sqrt{3}}{162\sqrt{\pi}} \left(\frac{Z}{a} \right)^{\frac{3}{2}} \rho(6 - \rho) \, e^{-\rho/3} \sin\theta \, e^{i\phi} \tag{3.41}$$

where $\rho = Zr/a$ and a is the Bohr radius $a = 0.529$ nm. The form of the orbital as the product of three separated functions $R(r)\Theta(\theta)\Phi(\phi)$ can clearly be seen.

So far we have considered the orbitals of the H atom in a representation defined by the three quantum numbers n, l and m_l. Since for given values of n and l there are $2l + 1$ different possibilities for m_l, and since the radial wave equation is independent of m_l, any combination of these $2l+1$ degenerate orbitals will also be a valid solution. The orbitals tabulated are in general complex functions, since the Φ function $e^{im\phi}$ has real and imaginary parts. It is common practice to take suitable real combinations of these functions as an alternative representation of the orbitals. For example the three possible $2p$ orbitals can be represented using the m quantum number by

$$
\begin{aligned}
2p_1 &= N_1 R_{2p}(r) \sin\theta \exp(i\phi) \\
2p_0 &= N_0 R_{2p}(r) \cos\theta \\
2p_{-1} &= N_1 R_{2p}(r) \sin\theta \exp(-i\phi)
\end{aligned}
\tag{3.42}
$$

or using the real combinations (NB $2p_0$ is already real)

$$
\begin{aligned}
2p_x &= \frac{1}{\sqrt{2}}(2p_1 + 2p_{-1}) = N_0 R_{2p}(r) \sin\theta \cos\phi \\
2p_z &= 2p_0 = N_0 R_{2p}(r) \cos\theta \\
2p_y &= \frac{1}{i\sqrt{2}}(2p_1 - 2p_{-1}) = N_0 R_{2p}(r) \sin\theta \sin\phi
\end{aligned}
\tag{3.43}
$$

Spatial probability distributions

One of the most important aspects of the interpretation of three dimensional wavefunctions such as atomic orbitals is the idea of probability density and the correct manipulation of the square modulus of the wavefunction to calculate the required probability.

As discussed in Chapter 1 $\psi^*\psi$ is a probability density. Thus the probability of finding the electron in a rectangular volume element of dimensions dx, dy and dz, at the point (x, y, z) is given by $\psi^*\psi\, dx\, dy\, dz$. This is the product of the probability density $\psi^*\psi$ and the infinitesimal volume $dx\, dy\, dz$. Similarly, in spherical polar coordinates, the probability of finding the electron in a volume element spanned by the dimensions dr, $d\theta$ and $d\phi$ is equal to the product of $\psi^*\psi$ with the volume of the element thus specified, which can be seen by reference to Fig. 3.6 to be $r^2 \sin\theta\, dr\, d\theta\, d\phi$.

The electron must exist somewhere in three-dimensional space, and so the orbital is normalized by integration over all three dimensions. For this purpose it is again most convenient to employ the representation in spherical polar coordinates.

$$
\int\int\int \psi^*\psi\, dx\, dy\, dz = \int_0^\infty \int_0^\pi \int_0^{2\pi} \psi^*\psi\, r^2 \sin\theta\, d\phi\, d\theta\, dr
$$

$$
= \int_0^\infty r^2 R^2 dr \int_0^\pi \sin\theta\, \Theta^2 d\theta \int_0^{2\pi} \Phi^*\Phi\, d\phi = 1
\tag{3.44}
$$

Since the normalization integral separates into the product of three one-dimensional integrals, it is possible to arrange that each of the functions $R(r)$, $\Theta(\theta)$ and $\Phi(\phi)$ is normalized separately, and this has been done in the tables.

We can now answer questions such as 'what is the probability density

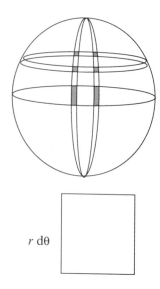

Fig. 3.6 The area element on the surface of a sphere is the rectangle of intersection of the two slices shown. The vertical sides are measured round great circles (lines of longitude with radius r) from θ to $\theta + d\theta$ and therefore have length $r\, d\theta$. The horizontal sides are measured from ϕ to $\phi + d\phi$ around a line of latitude, which is a circle of radius $r \sin\theta$, and they therefore have length $r \sin\theta d\phi$.

$r\, d\theta$

$r \sin\theta\, d\phi$

of the distance of the electron from the nucleus?'. The answer to this question can be found by writing the probability of finding the electron in the volume element $dr\, d\theta\, d\phi$ and then by integrating this probability over all possible values of θ and ϕ which will leave us with the over-all probability of finding the particle somewhere in the spherical shell between r and $r + dr$.

$$p_r(r)dr = r^2 R^2 dr \int_0^\pi \sin\theta\, \Theta^2 d\theta \int_0^{2\pi} \Phi^*\Phi\, d\phi \qquad (3.45)$$

Since the three functions R, Θ and Φ have been normalized separately in our tables, the two integrals in this equation are both equal to one, leaving the so-called radial distribution function

$$p_r(r) = r^2 R^2 \qquad (3.46)$$

In exactly the same way the probability densities for the two angles can be found.

$$p_\theta(\theta) = \sin\theta\, \Theta^2(\theta) \text{ and } p_\phi(\phi) = \Phi^*\Phi \qquad (3.47)$$

Sections through several different wavefunctions are depicted in Fig. 3.7, showing how the probability density varies as a function of space.

Because of the form of the wavefunction as a product, $R\Theta\Phi$, it is possible to identify three different types of node. A node is a place where ψ is zero. For ψ to be zero, at least one of the three functions R, Θ or Φ must be zero. If $R(r)$ is zero the wavefunction is said to have a radial node at r, and it is then zero everywhere on a sphere of radius r centred at the nucleus. Similarly, if $\Theta(\theta)$ is zero, then the wavefunction must be zero everywhere on a cone whose apex has half-angle θ. This is said to be an angular node. Both types of node can be seen in Fig. 3.7.

The form of the radial distribution function has profound chemical consequences. Although the $2s$ and the $2p$ orbitals have the same energy in the one-electron atom, this is no longer the case in many-electron atoms such as Li. These energy differences underlie the structure of the periodic table, since they determine the order in which the orbitals are filled. It is important to understand the reason for the removal of this degeneracy, which originates in the mutual repulsion energy between the electrons. At a simple level it can be seen from Figures 3.7 and 3.8 that the radial distribution function for the $2s$ orbital comes in two parts, separated by a radial node. One part is close to the nucleus, and the other part is further away. The $2p$ orbital has no such node, and its electron density is smaller than that of the $2s$ orbital close to the nucleus. In the Li atom the first two electrons occupy the $1s$ orbital, which forms the core. The outer electron is partially screened from the full nuclear charge by the core electrons. However, the $2s$ orbital is screened less effectively than the $2p$ orbital because it penetrates inside the core into a region where it experiences the full nuclear charge. The $2p$ orbital penetrates less effectively, and so lies at a higher energy than the $2s$ orbital. In consequence the $2s$ orbital is generally filled before the $2p$ orbital.

Of course the whole picture just presented is an oversimplification,

Problem 3.6.1. *Show that for an s orbital this expression of the radial distribution function is equivalent to the usual formula,* $4\pi r^2 \psi^2$.

An atomic orbital has a total of $n - 1$ nodes: l of these are angular nodes, and the remaining $n - l - 1$ are radial nodes.

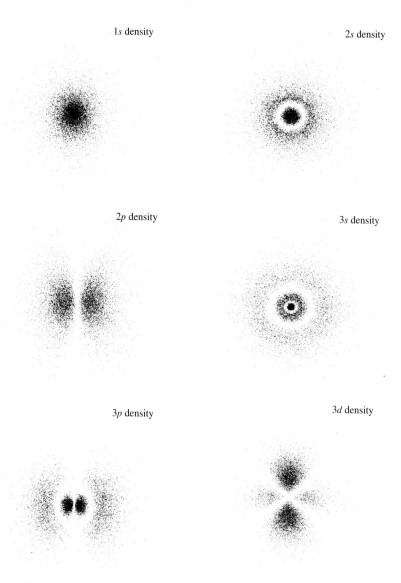

1*s* density

2*s* density

2*p* density

3*s* density

3*p* density

3*d* density

Fig. 3.7 Sections through the electron densities of different atomic orbitals of the H atom

since it relies on the picture in which each electron can be assigned to an orbital, the *orbital approximation* discussed in Section 2.7. However, the very fact that this approximate picture leads to a framework for rationalizing the structure of the periodic table illustrates its great utility, and explains why it is so widely used to describe the properties of multi-electron atoms.

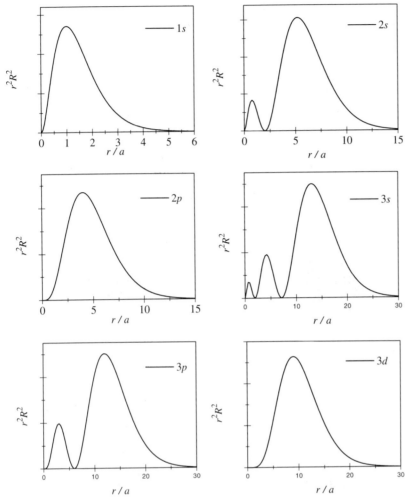

Fig. 3.8 Radial distribution functions for the H atom, $r^2 R^2$

Problem 3.6.2. *Use the tables to find for the H atom*

(a) *the radial and angular nodes of a $3p_0$ orbital;*

(b) *the radial distribution function, and hence the most probable distance from the nucleus of a 3d electron;*

(c) *the angular distribution function p_θ, and hence the most probable and least probable values of the angle θ for an electron in a $3d_0$ orbital. (The least probable values are at the angular nodes.)*

References

1. Landau, L. D. and Lifshitz, E. M. (1977). *Quantum mechanics.* Pergamon Press, Oxford.
2. Pauling, L. and Wilson, E.B. (1985). *Introduction to quantum mechanics.* Dover, New York.
3. Herzberg, G. (1950). *Spectra of diatomic molecules.* van Nostrand Reinhold, New York.
4. McQuarrie, D. A. (1983). *Quantum chemistry.* Oxford University Press, Oxford.
5. Bunker, P. R. (1979). *Molecular symmetry and spectroscopy.* Academic Press, San Diego.

4 Symmetries—group theory

The quantum mechanical problems we are most concerned with in chemistry involve finding the allowed energy levels of a molecule and the corresponding wavefunctions. Many such problems can be simplified considerably, and much insight can be gained by the judicious application of the ideas of symmetry. For example, we saw in Section 2.9 how to find the normal modes of vibration of a molecule, by diagonalizing the mass-weighted force constant matrix. However, as we shall see, each vibration has a particular symmetry, and only certain combinations of the coordinates have the correct symmetries. Use of these symmetry-adapted combinations brings the matrix into a form in which it is already partly diagonalized, making the remaining problem much simpler. In a similar way the simplest approximations to molecular orbitals are obtained by linear combination of atomic orbitals (i.e. by adding and subtracting suitably weighted atomic orbitals). The problem is to find the suitable weightings. However, each molecular orbital has a symmetry, and if we construct symmetry-adapted linear combinations, once again the size of the problem can be much reduced.

4.1 Symmetry operations

A symmetry operation is a transformation of a body in which the components of the body may be rearranged, but after which the body is not distinguishable from its original arrangement, including its orientation in space. In chemistry the bodies we consider are molecules, and the symmetry operations rearrange the constituent particles, but leave the molecule in an unchanged configuration. We can think of a symmetry transformation as being controlled by an operator. As usual we will distinguish an operator $\hat{R}$ from the operation it produces, R, by using a hat. A symmetry operator operates on a function to transform it into a different function, $\hat{R}|\psi> = |\phi>$. Symmetry operators are linear:

$$\hat{R}(a_1|1> + a_2|2>) = a_1\hat{R}|1> + a_2\hat{R}|2> \tag{4.1}$$

however, they are not necessarily hermitian because they do not represent observable quantities. We can define a hermitian conjugate operator by its effect on the conjugate of $|\psi>$, $<\hat{R}\psi| = <\psi|\hat{R}^+ = <\phi|$. In addition to linearity, symmetry operators have two basic properties:

1. They are unitary, meaning that $\hat{R}^+\hat{R} = \hat{E}$, the identity operator.

2. They commute with all operators corresponding to observables.

These mathematical properties represent the fact that no physical property of the molecule may be altered by a symmetry operation, as we now illustrate.

Firstly the overlap between two wavefunctions ψ_1 and ψ_2 is invariant to any symmetry operation. This arises from the unitary property of the symmetry operator $\hat{R}$,

$$< \hat{R}\psi_2|\hat{R}\psi_1 > = < \psi_2|\hat{R}^+\hat{R}|\psi_1 > = < \psi_2|\psi_1 > \qquad (4.2)$$

As a consequence of this result, symmetry operations do not alter normalization or orthogonality.

Secondly, the expectation value of any observable is invariant to any symmetry operation:

$$< \hat{R}\psi|\hat{A}|\hat{R}\psi > = < \psi|\hat{R}^+\hat{A}\hat{R}|\psi > = < \psi|\hat{R}^+\hat{R}\hat{A}|\psi > = < \psi|\hat{A}|\psi >$$
$$(4.3)$$

The second equality follows from the fact that $\hat{R}$ and $\hat{A}$ commute, and the final equality follows because $\hat{R}$ is unitary. Finally the eigenvalues of any operator are invariant under all symmetry operations. Suppose that ψ_a is an eigenfunction of an operator $\hat{A}$, representing an observable. Let us consider whether $\hat{R}\psi_a$ is also an eigenfunction of $\hat{A}$. Because $\hat{R}$ must commute with $\hat{A}$ we can write

$$\hat{A}\hat{R}\psi_a = \hat{R}\hat{A}\psi_a = \hat{R}a\psi_a = a\hat{R}\psi_a \qquad (4.4)$$

We have therefore proved that $\hat{R}\psi_a$ is also an eigenfunction of $\hat{A}$ with the same eigenvalue, a.

The relevance of symmetry to chemistry is a consequence of this final result, that application of a symmetry operation cannot alter the value of any observable, in particular the energy. Thus if the wavefunction ψ is an eigenfunction of $\hat{H}$ with energy E, then $\hat{R}\psi$ must also be an eigenfunction of $\hat{H}$ with the same energy. In other words, the Schrödinger equation is invariant to symmetry transformations. This invariance puts severe constraints on the possible forms of its solutions. For example, if an energy level of a molecule is nondegenerate, then the effect of all symmetry operators must be to transform the corresponding wavefunction into itself, or to alter it at most by the multiplication of a factor ± 1 (since the transformed function must still be normalized). The symmetry of a nondegenerate wavefunction is therefore determined by whether it is even or odd to each of the possible symmetry operations of the molecule.

By a similar argument it can be shown that symmetry operators do not alter the distance between two identifiable points (for example a bond length) or the angle between two vectors (for example a bond angle).

Symmetry operations must commute with all observables to ensure that no observable may be altered by the symmetry operation. If this were not the case a transformed molecule could be distinguished from the original by measuring this observable.

Types of symmetry operation

The symmetry operations of a molecule can be classified into several types: rotations, reflections, inversion, improper rotations. Each operation is associated with a *symmetry element*, for example rotations are associated with an axis and reflections with a mirror plane. The first stage in applying symmetry to a molecule is to identify its symmetry operations. We therefore give a brief discussion of each type of transformation in turn.

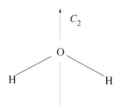

Fig. 4.1 The C_2 axis of the water molecule.

Fig. 4.2 The ammonia molecule viewed down the C_3 axis which passes through the N atom perpendicular to the plane of the paper and through the equilateral triangle of H atoms beneath.

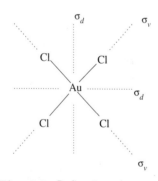

Fig. 4.3 Reflection planes of the planar $AuCl_4^-$ ion. The figure axis is C_4 and passes through the Au atom perpendicular to the plane of the paper, which is the σ_h plane. The σ_v planes contain the Au–Cl bonds and the σ_d planes bisect them.

Rotations

Rotation axes are classified by the number of successive rotations required to achieve a rotation of 2π radians. Thus a C_2 rotation axis is an axis for which rotation of π (180°) is a symmetry operation, as for the water molecule (see Fig. 4.1). Similarly for a C_3 axis, where rotation of 120° is a symmetry operation, as in the ammonia molecule (Fig. 4.2). The number n in a C_n axis is known as the *order* of the axis, and the highest order unique axis of a molecule, if there is one, is known as the *figure axis*.

Reflections

If the molecule contains a figure axis, then a mirror plane perpendicular to it is denoted σ_h (h for horizontal). Mirror planes which contain the figure axis are known as σ_v (v for vertical). Sometimes there are two nonequivalent types of plane containing the figure axis; these are then subdivided into σ_v and σ_d (dihedral). As an example, consider the square planar $AuCl_4^-$ ion where there is a C_4 figure axis. This ion has a horizontal plane, two vertical planes and two dihedral planes (see Fig. 4.3).

Inversion

The inversion transforms the coordinates $(x, y, z) \rightarrow (-x, -y, -z)$. The $AuCl_4^-$ ion has a centre of inversion, denoted i.

Improper rotations·

Improper rotations (S_n) consist of a rotation followed by a reflection in a plane perpendicular to the axis of rotation. In some cases neither the rotation nor the reflection alone is a symmetry operation, but the combination is. For example Fig. 4.4 shows the allene molecule ($CH_2=C=CH_2$) viewed from one end, down the C_2 axis; this axis is also a fourfold improper axis S_4. A rotation of 90° followed by a reflection in the plane of the paper transforms the molecule back into itself, but there is no C_4 axis, nor is the reflection plane a symmetry element of the molecule. Of course, in a molecule with a σ_h plane the figure axis C_n is automatically an S_n axis as well. An S_2 improper rotation is equivalent to an inversion.

4.2 Point groups

There is always at least one point in the molecule which is not moved by any of the symmetry operations. In the case of the $AuCl_4^-$ ion this invariant point is the central Au atom. For this reason, the collection of symmetry operations for a molecule are said to form a *point group*.

Group properties

A group is an abstract mathematical concept, defined as a collection of things which obey the following constraints:

1. The group must contain the identity operation E.

2. Each element must have an inverse such that $R^{-1}R = RR^{-1} = E$.

3. *Closure*: every combination of two elements must also be a member of the group, i.e. if R and S are members of the group, then so must be RS and SR.

4. *Associativity*: $(AB)C = A(BC)$.

There is no requirement that group members should commute with one another. In general they do not.

In practice the most important of these constraints is closure. We illustrate the property of closure by considering the symmetry operations of the ammonia molecule. This molecule has a C_3 axis, with two associated symmetry operations, C_3 and C_3^2, representing rotations by 120° and 240°, three σ_v planes and the identity operation. This group of operations is given the name C_{3v} (see the next subsection). Table 4.1 contains all possible combinations of these symmetry operations in pairs. It can be seen that every combination gives another element of the group, showing that the group is indeed closed.

The point group classification

Many different molecular shapes may have the same set of symmetry operations, and therefore belong to the same point group. For example, the point group C_{2v} consists of the identity, a C_2 operation and two distinct σ_v operations. This group of symmetry operations applies to H_2O, ClF_3, SF_4, *cis*- dichloroethene, *cis*-$[AuCl_2Br_2]^-$, *mer*-MX_3Y_3, and many other molecular geometries.

Point groups are named according to the combination of symmetry operations that forms the group. The most widely used system of nomenclature in chemistry is the Schoenfliess system, which we now summarize. (The total number of members of a group is known as its *order*.)

- C_n: a single C_n axis only, order $= n$. Examples: H_2O_2, 1,3 dichloroallene (CHCl:C:CHCl) (C_2).

- S_{2n}: a single S_{2n} axis only, order $= 2n$, note that $S_{2n}^2 = C_n$, and that $S_2 = i$. Example: CPh$_4$ (S_4).

- C_{nh}: a C_n axis and a horizontal plane, order $= 2n$ (n rotations, including E, and n improper rotations, including σ_h, and if n is even, i). Examples: *trans*-1,2 dichloroethene (C_{2h}), B(OH)$_3$ (C_{3h}).

- C_{nv}: a C_n axis with n vertical mirror planes, all containing the axis, order $= 2n$ (n rotations and n reflections). Examples: water, *cis*-1,2 dichloroethene, SF$_4$ (C_{2v}), ammonia (C_{3v}), CO ($C_{\infty v}$). A molecule of C_{nv} symmetry has the same symmetry operations as a pyramid whose base is an n-sided regular polygon. In the case of $n = 2$ this is an arrow-head, and in the case $n = \infty$ it is a cone.

Fig. 4.4 The allene molecule (CH$_2$=C=CH$_2$) viewed down the direction of the C=C=C bonds, which is both a C_2 axis and an S_4 axis. The full lines indicate H atoms bonded to the C atom above the plane of the paper and the dashed lines refer to H atoms below the plane of the paper.

The group properties can be memorized using the mnemonic INCA (Identity, Negation, Closure, Associativity).

Table 4.1 C_{3v} group multiplication table—BA

	First operation (A)				
E	C_3	C_3^2	σ_1	σ_2	σ_3
E	C_3	C_3^2	σ_1	σ_2	σ_3
C_3	C_3^2	E	σ_2	σ_3	σ_1
C_3^2	E	C_3	σ_3	σ_1	σ_2
σ_1	σ_3	σ_2	E	C_3^2	C_3
σ_2	σ_1	σ_3	C_3	E	C_3^2
σ_3	σ_2	σ_1	C_3^2	C_3	E

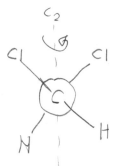

Fig. 4.5 A triangular prism with D_{3h} symmetry. The triangular faces are equilateral triangles. The H atoms of an ethane molecule in the eclipsed conformation occupy the vertices of the prism.

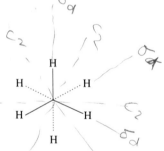

Fig. 4.6 An ethane molecule in the staggered conformation viewed down the C–C bond, which is both a C_3 axis and an S_6 axis. This geometry has D_{3d} symmetry.

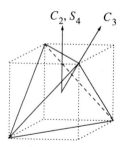

Fig. 4.7 A tetrahedron constructed on the vertices of a cube. One of the C_3 axes and one of the C_2–S_4 axes is shown.

- D_n: a C_n axis with n C_2 axes lying in a plane perpendicular to the figure axis, order $= 2n$ (n rotations about C_n and one rotation about each of the C_2). The C_2 axes can be hard to spot. Example: ethane in an arbitrary rotational conformation (D_3).

- D_{nh}: all the symmetry operations of D_n plus a horizontal plane perpendicular to the C_n axis and containing all the C_2 axes. A set of n vertical mirror planes automatically appears; each σ_v contains one of the C_2 axes. Order $= 4n$ (D_n plus n improper rotations (including σ_h) and n vertical reflections). Examples: ethene (D_{2h}), ethane (eclipsed), BCl$_3$ (D_{3h}), AuCl$_4^-$ (D_{4h}), ferrocene (eclipsed) (D_{5h}), benzene (D_{6h}), CO$_2$ ($D_{\infty h}$). This group is easy to spot since the molecules have the symmetry of a prism whose faces have n-fold rotational symmetry, or of a regular n-sided polygon (see Fig. 4.5).

- D_{nd}: As for D_n with the addition of n dihedral mirror planes which bisect the C_2 axes. Once the dihedral planes are added, the C_n axis also becomes an S_{2n} axis. Order $= 4n$ (D_n plus n σ_d reflections and n improper rotations not already counted as rotations about C_n). Examples: allene (D_{2d}), staggered ethane (D_{3d}), staggered ferrocene (D_{5d}). This group is also easy to spot—if the opposite faces in the prism of D_{nh} eclipse one another when viewed down the C_n axis, then they are staggered (exactly out of register) in D_{nd} (see Fig. 4.6).

- T_d: the symmetry operations of a tetrahedron. A tetrahedron is simple to recognize, but to apply group theory we need to locate all the symmetry operations. These are most easily represented by placing the vertices of the tetrahedron at the corners of a cube, as in Fig. 4.7. There are four C_3 axes passing through the body diagonals of the cube, and three mutually perpendicular C_2 axes joining the centres of opposite faces of the cube, giving a total of 12 rotations, including the identity. In addition there are six planes of symmetry, each of which contains one C_2 and bisects the other two. Thus each C_2 axis is also an S_4 axis, adding two improper rotations. The order of the group is therefore 24. Example: CH$_4$.

- O_h: the symmetry operations of a cube or an octahedron. The octahedron is more important in chemistry (except in the solid state) but the symmetry operations are easier to localize on the cube. The relationship between a cube and its inscribed octahedron is depicted in Fig. 4.8. The vertices of the octahedron are at the face centres of the cube. Again there are four C_3 axes through the body diagonals, and opposite faces of the cube are now joined by C_4 axes. Perpendicular to each C_4 axis there are two C_2 axes, joining opposite edges of the cube. There are therefore 24 rotational symmetry operations in the group. There is in addition a centre of symmetry, and so the C_3 axes automatically become S_6 axes, there are also six planes of symmetry passing through opposite edges, three more parallel to the faces of the cube, and the C_4 axes are converted into S_4 axes by the new horizontal planes, giving a total order of 48. Example: SF_6.

- I_h: the symmetry group of the regular icosahedron or dodecahedron. Contains six C_5 axes, ten C_3 axes and 15 C_2 axes, giving a total of 60 rotations, which doubles up to an order of 120 on adding the centre of inversion. It is not easy to depict the symmetry elements on a two-dimensional page, and we leave it as an exercise for the enterprising reader to locate them. This is the point group for the molecule C_{60}.

- R_3: the symmetry group of a sphere has to be treated differently since it contains an infinite number of rotations and reflections. However it is important for classifying spherically symmetrical systems, such as atoms, or the rotational wavefunctions of a diatomic molecule.

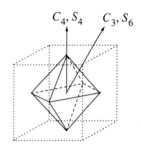

Fig. 4.8 An octahedron constructed inside a cube. One of the C_3–S_6 axes is shown coming out of the corner of the cube, and a C_2–S_4 axis coming through the centre of a face of the cube.

Finding the point group

The skill of finding and recognizing the point group of a molecule only comes with practice. However, to help the reader we include a flowchart in Fig. 4.9.

In searching for a point group it is useful to bear in mind that the most commonly found groups are C_{nv} and D_{nh}. These are easy to spot in practice because C_{nv} describes a pyramid whose base is a regular n-

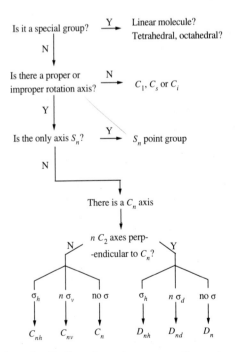

Fig. 4.9 Flowchart for finding the point group of a molecule.

sided polygon, and D_{nh} describes a prism whose ends are regular n-sided polygons.

Classes

Problem 4.2.1. *Show that if $\hat{A}$ is conjugate to $\hat{B}$ and $\hat{B}$ is conjugate to $\hat{C}$ then $\hat{A}$ is conjugate to $\hat{C}$.*

Two members of a group $\hat{A}$ and $\hat{B}$ are said to be *conjugate* if they are related by a transformation of the form (similarity transform)

$$\hat{B} = \hat{C}^{-1}\hat{A}\hat{C} \tag{4.5}$$

where $\hat{C}$ is also a member of the group (and conversely $\hat{A} = \hat{C}\hat{B}\hat{C}^{-1}$).

A collection of members of a group which are all conjugate to one another is known as a *class*. As an example we consider all possible combinations of the form $\hat{C}^{-1}\hat{A}\hat{C}$ for the C_{3v} group in Table 4.2, from which it can be seen that the identity is in a class of its own, the two rotation operations $\hat{C}_3$ and $\hat{C}_3^2$ form a class, and the three reflections $\hat{\sigma}_1$, $\hat{\sigma}_2$ and $\hat{\sigma}_3$ form a third class. The identity and the inversion commute with all other symmetry operations and therefore always form classes of their own, for example

$$\hat{C}\hat{E}\hat{C}^{-1} = \hat{E}\hat{C}\hat{C}^{-1} = \hat{E} \tag{4.6}$$

Table 4.2 C_{3v} similarity transform table—$B^{-1}AB$

		operation A				
B	E	C_3	C_3^2	σ_1	σ_2	σ_3
E	E	C_3	C_3^2	σ_1	σ_2	σ_3
C_3	E	C_3	C_3^2	σ_2	σ_3	σ_1
C_3^2	E	C_3	C_3^2	σ_3	σ_1	σ_2
σ_1	E	C_3^2	C_3	σ_1	σ_3	σ_2
σ_2	E	C_3^2	C_3	σ_3	σ_2	σ_1
σ_3	E	C_3^2	C_3	σ_2	σ_1	σ_3

4.3 Bases and representations

The purpose of using symmetry is to simplify the construction of molecular orbitals or vibrational modes by generating symmetry adapted combinations of the constituent atomic orbitals or atomic motions. Such simplifications can cut the computational work involved by orders of magnitude. For this purpose we do not need the whole of group theory, we simply concentrate on one aspect—the theory of *representations*.

Bases

A basis is a closed set of things, for example atomic orbitals, which are interconverted by the symmetry operations of the group.

Before discussing representations we need to understand the idea of a *basis*, which is a closed set of *things* that are transformed into one another by the operations of the group. The appropriate *things* are the building blocks in the application of interest. For example, if we are building molecular orbitals in the LCAO approximation, they are the constituent atomic orbitals. Or if we are interested in the vibrations of a molecule, then the basis chosen is a set of coordinates that can describe all possible displacements of the nuclei in space. The *dimension* of the basis is the number of things in it.

The most important feature of a basis is that it must be *closed*: i.e. it must be possible to express the effect of every symmetry operation of the group on every member of the basis as a linear combination of members of the basis, otherwise it is not a valid basis for the group. To take a very simple example, consider the 1s orbitals on the two H atoms in a water molecule (Fig. 4.10). The identity and reflection in the mirror plane which contains the molecule leave these orbitals unchanged, whereas the other two symmetry operations (C_2 and σ_v) interchange them. Thus, every symmetry operation operating on orbital $|1>$ yields either $|1>$ or $|2>$, and similarly for the transformations of $|2>$. We conclude that this set of two orbitals is a valid basis. However, orbital 1 is not a valid basis *on its own* because some operations transform it into orbital 2. As a second example we consider the set of orthogonal unit vectors based on the N atom in an ammonia molecule. The effect of each of the symmetry operations in the group is tabulated in Table 4.3. The threefold axis of the ammonia molecule has been chosen to be aligned with the threefold axis of the orthogonal coordinate system (the (1,1,1) direction) (see Fig. 4.11). Each symmetry operation interconverts the three unit vectors in some way, and so these vectors form a valid basis.

Once an appropriate basis has been found, the solution to the problem of interest is a linear combination of the basis functions. For example, if we are trying to construct a molecular orbital by the LCAO method, the MO can be expressed as a linear combinations of the AOs. The AOs make up the basis set $\{f_i\}$ and the MO can be expressed in the form

$$\psi = \sum_i a_i f_i \tag{4.7}$$

The notation can be made much more economical if we assemble the basis functions into a vector $\mathbf{f}$ and the coefficients into a second vector $\mathbf{a}$. The MO is then the dot product (scalar product) of the two vectors,

$$\psi = \mathbf{f}^T \mathbf{a} \tag{4.8}$$

(The superscript ^{T} designates the transpose, so that if $\mathbf{f}$ is a column vector, $\mathbf{f}^T$ is the corresponding row vector.) The vector of coefficients $\mathbf{a}$ specifies the solution precisely in terms of the basis $\mathbf{f}$. The problem of finding the solution is therefore reduced to that of finding the vector $\mathbf{a}$.

Representations

Once a valid basis has been chosen, we consider the effect of the symmetry operations on each of the basis functions. As discussed above, each symmetry operation transforms a basis function into a linear combination of the basis functions. Thus for example if the ith basis function is denoted f_i and R is a symmetry operator, then

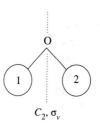

Fig. 4.10 H 1s orbitals in the water molecule.

Table 4.3 Displacement coordinates of the N atom in NH_3

	x	y	z
E	x	y	z
C_3	z	x	y
C_3^2	y	z	x
σ_1	x	z	y
σ_2	y	x	z
σ_3	z	y	x

Fig. 4.11 Displacement coordinates for the N atom in the ammonia molecule—the C_3 axis is aligned with the (1,1,1) direction of the axis system. The three axes point upwards out of the plane of the paper and the three N–H bonds point downwards.

The effect of a symmetry operation on the basis can be represented using a matrix. Each symmetry operation in the group has its own representative matrix, and the group of matrices is called a representation.

$$\hat{R}f_i = \sum_j f_j R_{ji} \tag{4.9}$$

where R_{ji} are the coefficients in the linear combination. If the basis functions are assembled into a row vector $\mathbf{f}^T$, then the coefficients R_{ji} form a matrix $\mathbf{R}$ which postmultiplies $\mathbf{f}^T$ to give exactly the same effect on the basis functions as the symmetry operator $\hat{R}$. In matrix notation

$$\hat{R}\mathbf{f}^T = \mathbf{f}^T \mathbf{R} \tag{4.10}$$

Each symmetry operation of the group can be represented in this way by a *representative matrix*, which is square and has dimension equal to the number of basis functions. This is also called the dimension of the representation.

So far we have only considered the effect of a symmetry operation on the basis functions $\mathbf{f}$. However, the molecular orbital or normal mode we want is not usually a single basis function, but a linear combination of them, which can be represented by a vector, as described in eqn 4.8. Given eqn 4.9 for the effect of a symmetry operation on a single basis function we have

$$\hat{R}\psi = \sum_i a_i \hat{R}f_i = \sum_{i,j} f_j R_{ji} a_i \tag{4.11}$$

or in the more compact matrix notation

$$\hat{R}\psi = \mathbf{f}^T \mathbf{R} \mathbf{a} \tag{4.12}$$

In other words the symmetry operation transforms the linear combination represented in the basis f by the vector $\mathbf{a}$ into the vector $\mathbf{Ra}$, relative to the same basis. The effect of two successive symmetry operations on the linear combination will tell us how the representative matrices should be combined to represent the effect of the combined symmetry operations. It is important to recollect that the notation $\hat{S}\hat{R}\psi$ signifies the operation of $\hat{R}$ first, followed by $\hat{S}$, because in general $\hat{R}$ and $\hat{S}$ will not commute. If the effect of $\hat{R}$ gives the vector $\mathbf{Ra}$, which we denote $\mathbf{b}$, then the effect of the second operation, $\hat{S}$, on $\mathbf{b}$ is given by $\mathbf{Sb}$, where $\mathbf{S}$ is the representative matrix of $\hat{S}$. Since matrix multiplication is associative we have,

$$\mathbf{Sb} = \mathbf{S(Ra)} = \mathbf{(SR)a} \tag{4.13}$$

In other words the combined symmetry operation $\hat{S}\hat{R}$ is represented in the basis $\mathbf{f}$ by the matrix product $\mathbf{SR}$. But the closure property of the group implies that this combination of symmetry operations must itself be a member of the group, whose effect on the vector $\mathbf{a}$ must be the same as that of the matrix $\mathbf{SR}$. The matrix product must therefore be the representative matrix of this operation. It follows from the preceding discussion that the representative matrices (with the operation of matrix multiplication to combine them) have the same multiplication table as the symmetry operators, and therefore form a group with exactly the same properties as the point group. They are said to form a *represen-*

representatives form a reppesentation of a group - may represent the effects of the symmetry operations on a particular basis

tation of the group because they represent the effects of the symmetry operations on a particular basis.

Every basis will generate an appropriate group of matrices with the same multiplication properties, i.e. every basis generates its own representation.

4.4 Reduction of representation

Representative matrices can become quite large. For example, if we take as a basis coordinates for displacing all the atoms in a methane molecule we obtain 15×15 matrices (three coordinates for each atom). The problem can be simplified considerably by a suitable coordinate transformation, and the resulting simplification is the main motivation for the use of group theory in chemistry.

Firstly we illustrate what we mean by another look at the example of the displacement coordinates of the N atom in the ammonia molecule. This time we align the axis system so that the z-axis lies along the figure axis of the molecule, and mirror plane 1 is the yz-axial plane (see Fig. 4.12). The effect of the possible symmetry operations is summarized in Table 4.4. Two features of these results should be noted:

1. The unit vector along the z axis is transformed into itself by all symmetry operations of C_{3v}, and is now a valid basis on its own for a representation whose representatives are 1×1 matrices (i.e. constants), all equal to 1.

2. The symmetry operations still mix the unit vectors in the x and y directions, which form a basis for a two-dimensional representation (consisting of 2×2 matrices).

The x and y vectors are related by symmetry in this example, because they are mixed by at least one symmetry operation in the group, however the z vector is clearly not mixed with either the x or y vectors by any symmetry operation. Effectively the three-dimensional space spanned by the x, y and z vectors has been separated into a one-dimensional sub-space (the z direction) and a two-dimensional sub-space (the xy plane). These sub-spaces are called invariant sub-spaces, which means that every symmetry operation transforms any vector in one of the sub-spaces into another vector in the same sub-space. For example, any symmetry operation will transform a vector in the xy plane into another vector in the xy plane. This separation into invariant sub-spaces, or *reduction of the representation* has been achieved simply by choosing the orientation of the coordinates carefully.

In this example we have reduced the representation by knowing the answer in advance. In general we do not know the answer in advance, but instead we use the mathematics of group theory, which gives a simple recipe for reducing representations.

The matrix representatives have the same multiplication table as the symmetry operations, and they therefore form a representation of the group.

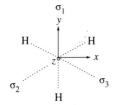

Fig. 4.12 Displacement coordinates for the N atom in the ammonia molecule—the C_3 axis is aligned with the z axis.

Table 4.4 Displacement coordinates of the N atom in NH_3

| | Coordinate | | |
	x	y	z
E	x	y	z
C_3	$\frac{-x+y\sqrt{3}}{2}$	$\frac{-x\sqrt{3}-y}{2}$	z
C_3^2	$\frac{-x-y\sqrt{3}}{2}$	$\frac{x\sqrt{3}-y}{2}$	z
σ_1	x	$-y$	z
σ_2	$\frac{-x-y\sqrt{3}}{2}$	$\frac{-x\sqrt{3}+y}{2}$	z
σ_3	$\frac{-x+y\sqrt{3}}{2}$	$\frac{x\sqrt{3}+y}{2}$	z

The process of finding a transformation to separate coordinates that are not related by symmetry is known as reducing the representation.

Transformation of basis

The two representations of the displacement coordinates of an ammonia molecule are obviously related. They differ only in the orientation of the axis system relative to the molecule. Reorienting the coordinate system is a linear transformation of the basis, which we now need to consider more generally.

Suppose we have two bases $\mathbf{f}$ and $\mathbf{g}$, which span the same space, and are related through the linear transformation $\mathbf{U}$, so that $\mathbf{f}^T = \mathbf{g}^T \mathbf{U}$;

$$f_i = \sum_j g_j U_{ji} \tag{4.14}$$

If $\mathbf{f}$ and $\mathbf{g}$ span the same space then $\mathbf{U}$ can be inverted so that $\mathbf{g}^T = \mathbf{f}^T \mathbf{U}^{-1}$. Our wavefunction ψ is a linear combination of these basis functions, and will be represented by a different vector in each basis. Similarly, every symmetry operation may be represented by a different matrix in each basis. If the wavefunction is represented by the vector $\mathbf{a}^{(f)}$ in basis $\mathbf{f}$ we need to discover how to find the vector $\mathbf{a}^{(g)}$ that represents it in basis $\mathbf{g}$. This is very simple because

$$\psi = \mathbf{f}^T \mathbf{a}^{(f)} = \mathbf{g}^T \mathbf{a}^{(g)} \tag{4.15}$$

Substituting for $\mathbf{f}^T$, we find

$$\psi = \mathbf{f}^T \mathbf{a}^{(f)} = \mathbf{g}^T \mathbf{U} \mathbf{a}^{(f)} \tag{4.16}$$

hence, we must have

$$\mathbf{a}^{(g)} = \mathbf{U} \mathbf{a}^{(f)} \text{ and } \mathbf{a}^{(f)} = \mathbf{U}^{-1} \mathbf{a}^{(g)} \tag{4.17}$$

Showing us that the matrix that transforms the *column vector* of the coefficients in a linear combination is the inverse of the matrix that transforms the *row vector* of the basis functions.

The second task is to find out how the transformation alters the representative matrix of a symmetry operator $\hat{R}$. Suppose that in basis $\mathbf{f}$ this matrix is $\mathbf{R}^{(f)}$, whereas in basis $\mathbf{g}$ it is $\mathbf{R}^{(g)}$. Then

$$\hat{R}\psi = \mathbf{f}^T \mathbf{R}^{(f)} \mathbf{a}^{(f)} = \mathbf{g}^T \mathbf{R}^{(g)} \mathbf{a}^{(g)} \tag{4.18}$$

Taking the first of these equalities and transforming the basis vector and the coefficient vector, as we have just discussed, we find

$$\mathbf{f}^T \mathbf{R}^{(f)} \mathbf{a}^{(f)} = \mathbf{g}^T \mathbf{U} \mathbf{R}^{(f)} \mathbf{U}^{-1} \mathbf{a}^{(g)} \tag{4.19}$$

Giving us the results

$$\mathbf{R}^{(g)} = \mathbf{U} \mathbf{R}^{(f)} \mathbf{U}^{-1} \text{ and } \mathbf{R}^{(f)} = \mathbf{U}^{-1} \mathbf{R}^{(g)} \mathbf{U} \tag{4.20}$$

This type of transformation is known as a *similarity transformation*. Representations which are related by a linear transformation of basis, like $\mathbf{f}$ and $\mathbf{g}$ are called *equivalent representations*.

Characters

To reduce a representation we need to find a transformation of the basis which will reduce the representative matrices to a block diagonal form. This can always be done for a single matrix, but the transformation we are looking for will reduce all the representative matrices to the same blocked form. Fortunately this job can be done without even constructing the representative matrices, all we actually need are their *characters*.

The character of a matrix (also known as its trace, or its spur) is the sum of its diagonal elements. We denote the character of a matrix with the greek letter χ. Thus

$$\chi(\mathbf{R}) = \sum_i R_{ii} \qquad (4.21)$$

When we transform from one basis to another the characters of the representative matrices do not change. This fact is easy to prove by reference to the similarity transformation of eqn 4.20.

$$\sum_i R_{ii}^{(g)} = \sum_{i,j,k} U_{ij} R_{jk}^{(f)} U_{ki}^{-1}$$

$$= \sum_{i,j,k} U_{ki}^{-1} U_{ij} R_{jk}^{(f)} = \sum_j R_{jj}^{(f)} \qquad (4.22)$$

because $\mathbf{U}^{-1}\mathbf{U}$ is the identity matrix. Hence $\chi(\mathbf{R}^{(g)}) = \chi(\mathbf{R}^{(f)})$, proving the assertion. This means that the representative matrix of $\hat{R}$ has the same character in all representations of the group equivalent to $\mathbf{f}$.

Another consequence of this observation is that in a given representation all matrices representing symmetry operations in the same *class* have the same character. This follows because the members of a class are related to one another by similarity transformations (see eqn 4.5).

This result makes things much easier. Instead of having to worry about the representative matrices for every symmetry operation, all we need to do is to find the character of one operation in each class; these characters are not altered when we transform the basis. And to find the character is simple—the diagonal elements of a matrix correspond to the components of the basis functions which are unaltered by the symmetry operation. In vector terms a diagonal element is the component of the transformed unit vector in the direction of the untransformed vector. For example, the identity operation leaves all the basis functions unaltered, so that each diagonal element of the representative matrix is one. When we calculate the character we count 1 for each basis function, giving the dimension of the representation.

The results of this subsection give us two other short-cuts:

1. Since every symmetry operation in a class has the same character, we only need to find this character for one member of each class. Published tables always group symmetry operations by class.

2. Since equivalent representations have the same set of characters

A blocked matrix has the form where the only nonzero elements come in small square blocks down the leading diagonal, as in

$$\begin{pmatrix} x & 0 & 0 \\ 0 & x & x \\ 0 & x & x \end{pmatrix}$$

which is blocked into a 1×1 and a 2×2 block.

Problem 4.4.1. *Check for each symmetry operation of the group C_{3v}, that the representative matrices in the two bases used to calculate Tables 4.3 and 4.4 have the same character.*

Problem 4.4.2. *Show, for the same set of matrices as the previous problem, that all matrices in the same class have the same character.*

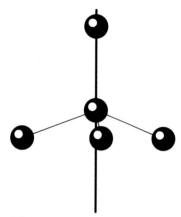

Fig. 4.13 Methane molecule, showing one of the four C_3 axes.

Problem 4.4.3. *Verify that the entries in the row of Table 4.5 labelled $\chi^{(\mathrm{H})}$ are correct. You may wish to refer to Fig. 4.7, which illustrates the symmetry operations of a tetrahedron.*

we can use different equivalent bases to work out characters for different classes. For example, it may be easier to orient the unit vectors of an orthogonal axis system differently when considering different symmetry operations.

To illustrate some of these points we work out the characters of the representation generated by the $1s$ orbitals on the H atoms of a methane molecule. For each symmetry operation, some of the orbitals are rearranged, and some stay where they were. The representative matrix describes how they are rearranged, but it only contains diagonal elements for orbitals whose positions are not altered. To find the character, we simply count the number of atoms which are not moved. For example, each C_3 axis passes through one of the atoms, which therefore does not move, but the other three atoms are cyclically permuted (Figure 4.13). Thus the character of the operation is 1, and the characters of the other seven operations in the class are immediately known to be 1 also. The characters of this representation for all the symmetry operations of the T_d point group will be found in Table 4.5 in the row labelled $\chi^{(\mathrm{H})}$. [The number apparently multiplying the symmetry operation at the top of each column indicates the number of members of the class.]

Table 4.5 Representations of the point group T_d

$\hat{R}$	E	$8C_3$	$3C_2$	$6S_4$	$6\sigma_d$
$\chi^{(\mathrm{H})}$	4	1	0	0	2
$\chi^{(x,y,z)}$	3	0	-1	-1	1

Problem 4.4.4. *Consider the representation generated by an orthogonal set of unit vectors on the C atom in a methane molecule. The atom stays where it is through all of the symmetry operations, but we need to count $+1$ in the character for a unit vector unchanged by an operation, -1 for a unit vector whose direction is reversed, and in general we need the component of the transformed unit vector in the direction of the original vector. Verify that the characters of this representation are as given in the row of Table 4.5 labelled $\chi^{(x,y,z)}$. [Hint: the axis system can be oriented differently for different operations—for the C_3 class align the C_3 axis along the (1,1,1) direction, for the the C_2 and S_4 classes choose the z axis to coincide with the symmetry axis, and for the σ_d let the mirror plane be the xy plane.]*

Irreducible representations

An irreducible representation, or irrep, is a representation whose basis functions cannot be separated by any linear transformation. They are inextricably mixed by the symmetry operations of the

We have already seen what it means to reduce a representation. The idea is to find a linear transformation of the basis which brings all the representatives into a block diagonal form. The transformed basis can then be divided into smaller subsets, each of which is a valid basis for a smaller representation. If such a transformation can be found, the representation is called *reducible*. If no such transformation exists, it is an *irreducible representation*, or *irrep* for short. An irrep is identified by a particular set of characters for the various symmetry operations of the group, which identifies its symmetry properties. For example, every group has an irrep whose characters are all unity, known as the *totally symmetric* irrep. Any function which is invariant to all the symmetry operations of the group is a basis for this irrep. For example the function obtained by adding the four H $1s$ orbitals in methane is a basis for the totally symmetric irrep of the point group T_d.

The dimension of an irrep is the number of basis functions required to generate it. The totally symmetric irrep always has a dimension of 1, but in point groups with a C_3 axis or higher, and a mirror plane or an inversion, there are irreps with dimensionality higher than one. The basis functions of such a *degenerate* irrep are inextricably transformed into one another by the operations of the group. No linear transformation exists which can separate them from one another. It is always possible to find a transformation which will diagonalize *some* of the representative matrices, but the representation is irreducible if there is no transformation which will diagonalize *all* of them at once. For example, consider the (x, y) basis in the ammonia molecule. If the axes are oriented so that the z axis is parallel to the figure axis and H atom 1 lies in the yz plane (see Fig. 4.12) the representative matrix for the σ_v operation in this plane is

$$\begin{pmatrix} -1 & 0 \\ 0 & 1 \end{pmatrix}$$

group, and are therefore symmetry related. The space spanned by the basis functions does not contain any invariant sub-space.

The representative matrix for the reflection in the plane containing atom 2 is not diagonal in this basis. It can be diagonalized by orienting the coordinate axes so that atom 2 lies in the yz plane, but once this transformation is made, the representative matrix of σ_1 will no longer be diagonal. It is not possible to find a transformation that diagonalizes both matrices at once, and so the representation is irreducible.

Symmetry labels

Each irrep has a different set of characters for the symmetry operations of the group, and is given a label which helps to indicate its symmetry properties. We now describe these briefly. A symmetry label generally consists of a letter, a subscript and/or a superscript. The letter indicates the dimension of the irrep, and the superscript or subscript denotes transformation properties for particular types of symmetry operation:

1. The letters A or B indicate a one-dimensional irrep. E denotes a two-dimensional irrep and T (or F) a three-dimensional irrep. A irreps are symmetric with respect to rotation about the figure axis, or all the axes of highest order if there is no unique figure axis, whereas B is antisymmetric with respect to this rotation (around at least one of the axes of highest order), which therefore has a character of -1.

2. Irreps in point groups with a horizontal mirror plane σ_h, but no

inversion, are labelled with a prime $'$ (or double prime $''$) indicating symmetry (or antisymmetry) with respect to the σ_h.

3. Irreps in point groups with inversion symmetry are labelled with subscripts g (or u) indicating symmetry (or antisymmetry) with respect to inversion.

4. Numerical subscripts indicate the presence of more than one irrep with the same classification, which differ in symmetry with respect to some other operation, e.g. rotation about a C_2 axis.

Reducing the representation

A typical representation is made up of a number of irreps of each symmetry. Reducing the representation tells us how many different irreps of each symmetry are contained in the representation.

Each point group has only a small number of possible irreps. In consequence there may be several independent linear combinations of the basis functions with the same symmetry. Group theory enables us to find how many independent combinations there are for each irrep. In more colourful terminology the representation contains a certain number of irreps of each type. The reduction of the representation tells us how many irreps of each type are hiding in it. It will also tell us how to find the correct symmetry-adapted combinations of the basis functions to generate each irrep.

It is conventional to label a representation with the greek letter Γ and the ith irreducible representation $\Gamma^{(i)}$. Then Γ can be expressed in terms of its constituent irreps by

$$\Gamma = \sum_i a_i \Gamma^{(i)} \tag{4.23}$$

which means that the basis for the representation Γ contains a_i independent linear combinations of symmetry $\Gamma^{(i)}$. In other words it is possible to find a transformation which separates the basis into subsets that are not interrelated by symmetry. Each subset will generate one block in all the representative matrices of the group. a_i of these subsets will have the symmetry properties of the irrep $\Gamma^{(i)}$. Reducing the representation amounts to finding the coefficients a_i, which must all be positive integers (or zero). After reduction, all the representative matrices will be in block diagonal form. Since the reduction will be effected by a similarity transform, Γ and its reduction are equivalent, and so the character of any particular representative matrix is unaltered by the reduction. The character of the block-diagonal form is simply the sum of the diagonal elements of all the elementary blocks down the diagonal. It follows that for each symmetry operation R

$$\chi(R) = \sum_i a_i \chi^{(i)}(R) \tag{4.24}$$

where $\chi^{(i)}$ is the character of R in irrep i. If we can find a set of coefficients a_i which satisfies this equation for each symmetry operation then we shall have succeeded in reducing the representation.

The reduction formula

Group theory gives us a simple formula for reducing a representation to its constituent irreps The proof of this formula is beyond the scope of this book. The interested reader is referred to an advanced text on group theory.[2,7,8] The major result is an orthogonality theorem for the elements of the representative matrices $\mathbf{R}^{(i)}$ and $\mathbf{R}^{(j)}$. In a group of order h the theorem reads

$$\sum_R R_{kl}^{(i)} R_{mn}^{(j)} = \frac{h}{n_i}\delta_{ij}\delta_{km}\delta_{ln} \qquad (4.25)$$

where n_i is the dimension of irrep i, $\delta_{ij} = 1$ if $i = j$ and 0 otherwise, and the summation extends over all the symmetry operations of the group. This summation is identically zero unless the two irreps are the same and the indices on the matrix elements are the same.

Since we are interested in the characters of the representative matrices, we apply this relation to the diagonal elements,

$$\sum_R R_{kk}^{(i)} R_{mm}^{(j)} = \frac{h}{n_i}\delta_{ij}\delta_{km} \qquad (4.26)$$

If we now sum over all possible values of k and m we find that the characters from different irreps are orthogonal.

$$\sum_R \chi^{(i)}(R)\chi^{(j)}(R) = h\delta_{ij} \qquad (4.27)$$

where $\chi^{(i)}$ is the character of $\mathbf{R}$ in the irrep $\Gamma^{(i)}$. This result can be used to reduce a representation, since we can now think of the characters for our reducible representation, Γ, as a vector, made up from orthogonal components in the directions of each irrep. The coefficient a_i is the projection of the vector in the direction corresponding to the irrep $\Gamma^{(i)}$. Hence we obtain the reduction formula

$$a_i = \frac{1}{h}\sum_R \chi^{(i)}(R)\chi(R) \qquad (4.28)$$

To apply this reduction formula we need to know the set of characters for the representation we are reducing, and also a set of characters for each irrep of the point group, which can be found in published tables.[9]

Another consequence of the orthogonality theorem is sometimes useful. Taking the case $i = j$ in eqn 4.27,

$$\sum_R |\chi^{(i)}(R)|^2 = h \qquad (4.29)$$

This equation is true for each irrep and it does not hold if the representation is reducible. It can therefore be used to check whether a particular representation is reducible. As an example we take the representations in Table 4.5 and check them for reducibility. The character table for the T_d point group is given in Table 4.6.

The formula for reducing a representation requires only the characters of the representations concerned. This formula is an example of a general result called an orthogonality theorem.

When applying the reduction formula or any other summation over the symmetry operations we must remember that published character tables list classes rather than operations, because each symmetry operation in the class has the same character. Thus we have to weight each term in the summation according to the number of elements of each class.

Table 4.6 Character table for the point group T_d

$\hat{R}$	E	$8C_3$	$3C_2$	$6S_4$	$6\sigma_d$			
A_1	1	1	1	1	1		$x^2 + y^2 + z^2$	
A_2	1	1	1	-1	-1			
E	2	-1	2	0	0		$(2z^2 - x^2 - y^2, x^2 - y^2)$	
T_1	3	0	-1	1	-1	(R_x, R_y, R_z)		
T_2	3	0	-1	-1	1	(x, y, z)		(xy, yz, xz)
$\chi^{(\mathrm{H})}$	4	1	0	0	2			
$\chi^{(x,y,z)}$	3	0	-1	-1	1			

For $\Gamma^{(\mathrm{H})}$ we find that the sum of the squares of the characters is $1 \times 4^2 + 8 \times 1^2 + 3 \times 0^2 + 6 \times 0^2 + 6 \times 2^2 = 48$, whereas for $\Gamma^{(x,y,z)}$ we get 24 which is the order of the group. We therefore conclude that $\Gamma^{(x,y,z)}$ is an irrep, whereas $\Gamma^{(\mathrm{H})}$ is reducible. A glance at Table 4.6 tells us that $\Gamma^{(x,y,z)} = T_2$. We now use the reduction formula to reduce $\Gamma^{(\mathrm{H})}$.

$$
\begin{aligned}
A_1 &: (1.1.4 + 8.1.1 + 3.1.0 + 6.1.0 + 6.1.2)/24 = 1 \\
A_2 &: (1.1.4 + 8.1.1 + 3.1.0 - 6.1.0 - 6.1.2)/24 = 0 \\
E &: (1.2.4 - 8.1.1 + 3.2.0 + 6.0.0 + 6.0.2)/24 = 0 \\
T_1 &: (1.3.4 + 8.0.1 - 3.1.0 + 6.1.0 - 6.1.2)/24 = 0 \\
T_2 &: (1.3.4 + 8.0.1 - 3.1.0 - 6.1.0 + 6.1.2)/24 = 1
\end{aligned}
\tag{4.30}
$$

Hence $\Gamma^{(\mathrm{H})} = A_1 + T_2$.

4.5 Energy levels, degeneracies and irreps

Every electronic or vibrational energy level of a molecule has an associated wavefunction which is a basis for an irrep of the molecular point group. Thus every energy level can be given a symmetry label, which gives the transformation properties of the corresponding wavefunction.

Symmetry has far-reaching consequences in quantum mechanics. Every electronic or vibrational wavefunction is a basis for an irrep of the molecular point group, and so the corresponding energy level can be labelled with the symmetry label of the irrep. Furthermore, the dimension of the irrep is the degeneracy of the energy level. We can understand the relationship between symmetry and energy levels in the following way. The action of a symmetry operation leaves the molecule in an arrangement indistinguishable from its original configuration. The energy of the molecule is therefore not altered. In more formal terms, every symmetry operation commutes with the Hamiltonian, and so

$$
\hat{H}\hat{R}\psi = \hat{R}\hat{H}\psi = E\hat{R}\psi
\tag{4.31}
$$

Thus each symmetry operation must transform the molecular wavefunction into a wavefunction of the same energy. Every symmetry operation must transform this wavefunction into itself (or minus itself). Such a function is therefore a basis for a one-dimensional irrep, whose characters are ±1 for all symmetry operations, and cannot be related by symmetry to any other wavefunction. The corollary to this argument is that if a wavefunction generates a one-dimensional irrep, there is no reason from symmetry for it to be degenerate with any other wavefunction. In general it will be nondegenerate, apart perhaps from an accidental degeneracy. On the other hand, a two-dimensional irrep requires two linearly independent basis functions. These two functions are interconverted, or mixed, by some of the symmetry operations. It is not possible to find any transformation which separates them; there must always be some symmetry operation which mixes them. Thus two independent wavefunctions which generate a two-dimensional irrep must be degenerate. Similarly a set of three wavefunctions which generates an irrep of dimension three must be triply degenerate.

It is easy to see by reference to a set of character tables that it is not possible for a molecule to have an orbital or vibrational degeneracy unless its point group contains at least a C_3 rotation axis, and it is not possible to have a triple degeneracy except in point groups with tetrahedral symmetry or higher. Thus we are not likely to find any degeneracies (except by accident) in the water molecule, whose figure axis is C_2. In the ammonia molecule we expect a maximum degeneracy of 2, since there is a unique C_3 axis, and we should not be surprised to find triply degenerate molecular orbitals and vibrations in the methane molecule.

4.6 Projection operators

We have seen how to decompose a representation into its constituent irreps, and we have investigated the consequences of this decomposition for the energy levels of a molecule. To complete the analysis we also need to find the correct symmetry-adapted combinations of the original basis functions that are appropriate to generate each irrep. These combinations can be found using a *projection operator* for each irrep i, defined by

A projection operator enables us to find the correct symmetry-adapted combinations of basis functions for each irrep present in the representation.

$$\hat{P}^{(i)} = \sum_R \chi^{(i)}(R)\hat{R} \tag{4.32}$$

where, as usual, the sum covers all symmetry operations of the group. The effect of a symmetry operation on a basis function is to transform it into some combination of the basis functions. The projection operator will also give a combination of the basis functions, but the combination projected out will belong to the irrep $\Gamma^{(i)}$. If we apply the projection operator $\hat{P}^{(i)}$ to each element of the basis set in turn, a complete basis for irrep i will be projected out. Of course, the original basis frequently contains far more functions than are needed to generate an irrep. The projection operator will then give several copies of the same basis function, or of linearly dependent basis functions.

As an example we consider the representation generated by the four H $1s$ orbitals in the methane molecule. We have already shown that this representation can be reduced to $A_1 + T_2$, and now we apply the projection operator method to find combinations of the four basis functions that act as bases for these irreps. To start with we make a table of the effect of all the symmetry operations of the group on the four basis functions. This will be found in Table 4.7. We then apply the projection operator for the irrep A_1 to each of the four basis functions in turn. When applied to each basis function this projection operator gives the

Table 4.7 Effect of the symmetry operations on basis H in T_d

E	H_1	H_2	H_3	H_4
$C_3(1)$	H_1	H_4	H_2	H_3
$C_3^2(1)$	H_1	H_3	H_4	H_2
$C_3(2)$	H_3	H_2	H_4	H_1
$C_3^2(2)$	H_4	H_2	H_1	H_3
$C_3(3)$	H_2	H_4	H_3	H_1
$C_3^2(3)$	H_4	H_1	H_3	H_2
$C_3(4)$	H_2	H_3	H_1	H_4
$C_3^2(4)$	H_3	H_1	H_2	H_4
$C_2(x)$	H_4	H_3	H_2	H_1
$C_2(y)$	H_3	H_4	H_1	H_2
$C_2(z)$	H_2	H_1	H_4	H_3
$S_4(x)$	H_3	H_1	H_4	H_2
$S_4^3(x)$	H_2	H_4	H_1	H_3
$S_4(y)$	H_2	H_3	H_4	H_1
$S_4^3(y)$	H_4	H_1	H_2	H_3
$S_4(z)$	H_3	H_4	H_2	H_1
$S_4^3(z)$	H_4	H_3	H_1	H_2
σ_{12}	H_1	H_2	H_4	H_3
σ_{13}	H_1	H_4	H_3	H_2
σ_{14}	H_1	H_3	H_2	H_4
σ_{23}	H_4	H_2	H_3	H_1
σ_{24}	H_3	H_2	H_1	H_4
σ_{34}	H_2	H_1	H_3	H_4

Problem 4.6.2. *It is usual (but not necessary) to find orthogonal combinations of these three functions. Check that the combinations*

$g_1 = 3H_1 - (H_2 + H_3 + H_4)$

$g_2 = 2H_2 - (H_3 + H_4)$

$g_3 = H_3 - H_4$ (4.33)

fit the bill in this example.

Group theory also tells us how the symmetry of a wavefunction is altered if the molecule is distorted in some way, giving it a lower symmetry. In particular it is simple to predict how a degenerate energy level will split in the distorted molecule.

same result, $6(H_1 + H_2 + H_3 + H_4)$. We therefore conclude that the sum of the four original basis functions is a basis for the irrep A_1.

Problem 4.6.1. *Verify, by considering the effect of each symmetry operator, that this combination is a basis for A_1.*

Application of the projection operator for T_2 to each function in turn gives

$$\hat{P}^{(T_2)}H_1 = 6H_1 - 2(H_2 + H_3 + H_4)$$
$$\hat{P}^{(T_2)}H_2 = 6H_2 - 2(H_1 + H_3 + H_4)$$
$$\hat{P}^{(T_2)}H_3 = 6H_3 - 2(H_1 + H_2 + H_4)$$
$$\hat{P}^{(T_2)}H_4 = 6H_4 - 2(H_1 + H_2 + H_3)$$

The first three of these are linearly independent of one another, but the fourth can be expressed in terms of the other three:

$$\hat{P}^{(T_2)}H_1 + \hat{P}^{(T_2)}H_2 + \hat{P}^{(T_2)}H_3 = -\hat{P}^{(T_2)}H_4$$

Three independent functions are needed to generate a three-dimensional irrep such as T_2, thus the projection operator method has given us a complete basis for generating T_2 (see Problem 4.6.2).

Group theory therefore gives us mechanisms, not only for decomposing a representation into its constituent irreps (the reduction formula), but also for constructing the correct symmetry-adapted linear combinations of the basis functions to reduce the representation. These combinations are used to construct the wavefunctions of the system.

4.7 Descent of symmetry

Another important use of group theory in chemistry is to predict what will happen to a particular degeneracy if a molecule or atom is distorted, removing some of its symmetry. The most familiar example is of an atom with five degenerate d orbitals, introduced into an environment with octahedral symmetry, which are split into one triply degenerate group (T_{2g} symmetry) and one doubly degenerate group (E_g symmetry). The irreps of the distorted system can be determined from the irreps of the undistorted system, because the distortion removes some of the symmetry operations, but retains others, which form a *subgroup* of the original point group. Because all the symmetry operations of the subgroup are also elements of the original group, a set of functions which was a basis for an irrep of the original group will still be a basis for some representation of the subgroup, but this may now be a reducible representation.

As an example we consider a CH_4 molecule stretched along one of its C_3 axes (passing through H atom 1). The point group of the distorted molecule is C_{3v} (Table 4.8), which is a subgroup of T_d, since all the symmetry operations of the distorted molecule are also symmetry operations of the original molecule. In particular let us consider the three functions in eqn 4.33, that form a basis for the T_2 representation of T_d. The characters of the symmetry operations of C_{3v} on these basis functions can be read straight from Table 4.6. The C_3 axis is one of the four C_3 axes in T_d, so that the character is 0. Similarly the three σ_v planes of the distorted molecule are among the six σ_d planes of T_d, and so the character for each of these operations is 1.

Thus under this type of trigonal distortion all triply degenerate energy levels of symmetry T_2 in a tetrahedral molecule split into a doubly degenerate pair (of symmetry E) and a nondegenerate wavefunction (A_1). It is easily verified that g_1 is a basis for the nondegenerate irrep A_1, and that g_2 and g_3 are degenerate.

4.8 Vanishing integrals

A great deal of the work involved in investigating the quantum mechanical properties of molecules consists of the calculation of matrix elements, which are generally integrals of the form $\int \phi(\mathbf{r})d\mathbf{r}$. However, many of the possible integrals in a particular application turn out to be zero. We can use group theory to tell us when an integral of this form must be identically zero. The method is a generalization of the idea of even and odd functions. We know that if we take an odd function, for which $\phi(-x) = -\phi(x)$, then its integral over all space is zero, because the integrals from $-\infty$ to 0 and from 0 to ∞ are equal and opposite in sign. In quantum mechanics the integrand is usually more complicated than this, since it covers a multi-dimensional space, but the same idea can be used. For example, if we can find a symmetry transformation which transforms the integrand into minus itself then the integral must be identically zero, because the function is odd with respect to this symmetry transformation: for every region where the integrand is positive, there is a symmetry-related region where it is negative.

The idea can be generalized further. If the integral represents a physical quantity, then no symmetry operation can alter its value:

$$\int \hat{R}\phi \, d\mathbf{r} = \int \phi d\mathbf{r} \tag{4.34}$$

We start by considering the case where ϕ is a basis function for irrep f. Let us rename it f_k for the moment. If we apply each symmetry operation in turn and sum, then the result must be h (the order of the group) times the value of the integral:

$$\sum_R \int \hat{R} f_k d\mathbf{r} = h \int f_k d\mathbf{r} \tag{4.35}$$

because each term in the summation is equal. We can express the operation of $\hat{R}$ in terms of its representative matrix $\mathbf{R}^{(f)}$, giving

$$\sum_l \sum_R R_{kl}^{(f)} \int f_l d\mathbf{r} = h \int f_k d\mathbf{r} \tag{4.36}$$

But we can now use the orthogonality eqn 4.25, because in the totally

Table 4.8 C_{3v} group character table

	E	$2C_3$	$3\sigma_v$
A_1	1	1	1
B_1	1	1	-1
E	2	-1	0

Problem 4.7.1.
Decompose this representation using the reduction formula, and show that it is equal to $A_1 + E$.

A quantum mechanical integral, such as an expectation value, must be identically zero unless the integrand spans the totally symmetric irrep of the point group.

symmetric irrep, $\Gamma^{(1)}$, all the representative matrices are equal to unity, hence

$$h \int f_k \mathrm{d}\mathbf{r} = \sum_l \sum_R R_{kl}^{(f)} R_{11}^{(1)} \int f_l \mathrm{d}\mathbf{r}$$

$$= \sum_l \frac{h}{n_1} \delta_{k1} \delta_{l1} \delta_{f1} \int f_l \mathrm{d}\mathbf{r} = h \delta_{i1} \int f_1 \mathrm{d}\mathbf{r} \quad .$$

The integral is therefore zero unless the integrand is a basis for the totally symmetric irrep. This is the result we were looking for; if the integrand is not a basis function for this irrep then it is in some sense an odd function and its integral over all space is necessarily zero. If the integrand ϕ is a basis function for a reducible representation, then following reduction of the representation, it can always be expressed as a linear combination of basis functions for irreps. The only terms in this sum which contribute to the integral are those which generate the totally symmetric irrep. In other words, if the reduced representation does not include irrep $\Gamma^{(1)}$ then the integral must be zero by symmetry. We add one word of caution at this point. Group theory only gives us a sufficient condition for the integral to vanish. An integral can still be zero for other reasons.

4.9 Direct products

When wavefunctions are multiplied together, or acted upon by an operator, the component parts may have different symmetries. The direct product representation tells us how these symmetries are combined in the final product.

We have seen in the last section how to use symmetry to tell whether an integral must vanish for symmetry reasons. In this section we complete our rapid tour of group theory by investigating the symmetry of the integrand in a matrix element. Matrix elements in quantum mechanics are generally of the form $< g_j |\hat{\Omega}| f_i >$, where g_j and f_i are basis functions for irreps $\Gamma^{(g)}$ and $\Gamma^{(f)}$, and the operator $\hat{\Omega}$, which is also a function of the space coordinates, transforms according to irrep $\Gamma^{(\Omega)}$. To determine whether the integral vanishes we need the representation generated by the entire integrand. The new feature we need for this purpose is the *direct product*. Let us start by considering a simpler matrix element of the form $< g_j | f_i >$. The integrand is the product function $g_j f_i$, and so we consider the representation generated by using the set of all such product functions as a basis. This is called the direct product representation of $\Gamma^{(f)}$ and $\Gamma^{(g)}$. The effect of a symmetry operation R on this product function is easy to determine, since f_i is transformed to $\hat{R} f_i$, g_j to $\hat{R} g_j$, and the two transformed functions are multiplied together. In terms of the representative matrices $\mathbf{R}^{(f)}$ and $\mathbf{R}^{(g)}$:

$$\hat{R} g_j f_i = \sum_{k,l} R_{jk}^{(g)} R_{il}^{(f)} g_k f_l \qquad (4.37)$$

The diagonal element of this double summation is the coefficient of the original function $g_j f_i$ on the right hand side, i.e. the coefficient $R_{jj}^{(f)} R_{ii}^{(g)}$. The character of R in the direct product representation is the sum of these diagonal elements over all possible values of i and j, which is therefore

$$\chi^{(fg)}(R) = \chi^{(g)}(R)\chi^{(f)}(R) \qquad (4.38)$$

The character of the symmetry operation R in the direct product representation is obtained by multiplying together the characters of R in the two constituent representations, $\Gamma^{(g)}$ and $\Gamma^{(f)}$.

We can find general rules for direct products, for example:

1. If the representations $\Gamma^{(g)}$ and $\Gamma^{(f)}$ are both degenerate, then the direct product representation will be reducible. If one of them is of dimension one then the direct product representation will be irreducible.

2. If $\Gamma^{(f)}$ is totally symmetric, then its characters are all equal to one. In this case $\Gamma^{(g)}$ and $\Gamma^{(fg)}$ are equivalent (they have the same character for each symmetry operation).

3. The direct products of three representations, $\Gamma^{(g)} \times \Gamma^{(\Omega)} \times \Gamma^{(f)}$, can be found by multiplying the three sets of characters together.

4. If the direct product representation does not contain the totally symmetric irrep then all matrix elements must be zero.

5. The direct product of an irrep with itself always contains the totally symmetric irrep $\Gamma^{(1)}$. The direct product of two different irreps never does.

The last two rules have an interesting consequence. The integral $< g_j|f_i >$ must be zero unless g_j and f_i are bases for the same irrep. In other words, wavefunctions with different symmetries must be orthogonal.

A particularly important role is played by the direct product of a representation with itself, since the last result tells us that the matrix element $< g_j|f_i >$ must be zero unless $\Gamma^{(g)}$ and $\Gamma^{(f)}$ are identical, as we now prove: the direct product representation has characters $\chi^{(g)}(R)\chi^{(f)}(R)$. In order to find whether or not this representation contains $\Gamma^{(1)}$ we apply the reduction formula, eqn 4.28

$$a_1 = \sum_R \chi^{(fg)}(R)\chi^{(1)}(R) = \sum_R \chi^{(f)}(R)\chi^{(g)}(R) = h\delta_{fg}$$

The second line follows from the formula for the direct product characters and from the fact that all characters in $\Gamma^{(1)}$ are unity. The third line follows from eqn 4.27.

Symmetric and antisymmetric products

The symmetries which arise when functions of the same symmetry are multiplied together are particularly important in chemistry. Suppose that $\{f_i\}$ and $\{g_j\}$ are different sets of basis functions for the *same* irrep, $\Gamma^{(f)}$. A partial reduction of the direct product representation can be obtained by considering separately the symmetrized product functions $f_i g_j + g_i f_j$ and the antisymmetrized product functions $f_i g_j - g_i f_j$.

Each of these sets of functions is separately a basis for a representation of the group:

$$\hat{R}(g_j f_i \pm g_i f_j) = \sum_{k,l} R_{kj}^{(f)} R_{li}^{(f)} (g_k f_l \pm g_l f_k)$$

$$= \frac{1}{2} \sum_{k,l} (R_{kj}^{(f)} R_{li}^{(f)} \pm R_{lj}^{(f)} R_{ki}^{(f)})(g_k f_l \pm g_l f_k) \qquad (4.39)$$

the second line follows by considering what happens if the indices k and l are reversed. We can also work out the character of these two representations: the diagonal element in the summation is given by

$$\frac{1}{2}(R_{jj}^{(f)} R_{ii}^{(f)} \pm R_{ji}^{(f)} R_{ij}^{(f)})(g_j f_i \pm g_i f_j) \qquad (4.40)$$

Summing over all possible values of i and j, and recognizing the second term in the summation as the matrix product of $\mathbf{R}^{(f)}$ with itself we obtain

$$\chi^{(\pm)}(R) = \frac{1}{2}(\chi^{(f)}(R)^2 \pm \chi^{(f)}(R^2)) \qquad (4.41)$$

If the dimension of $\Gamma^{(f)}$ is n then the direct product representation has dimension n^2, and this splits into a symmetrized representation of dimension $\frac{1}{2}n(n+1)$ and an antisymmetrized representation of dimension $\frac{1}{2}n(n-1)$. The combinations with $i = j$ are missing from the antisymmetrized representation because $f_i g_i - g_i f_i$ is identically zero.

As an example we find the direct product representation $T_2 \times T_2$ in the point group T_d and divide it into its symmetric and antisymmetric parts. The relevant characters are given in Table 4.10. The characters of the direct product representation are simply the squares of the characters of T_2. The characters of R^2 can be found easily since C_3^2 is still in the C_3 class, $C_2^2 = \sigma_d^2 = E$ and $S_4^2 = C_2$. These representations are reduced by the usual method to give

$$\Gamma^{(+)} = A_1 + E + T_2 \text{ and } \Gamma^{(-)} = T_1$$

Of course, since $\Gamma^{(-)}$ is an irrep, it can be identified immediately from Table 4.6.

Sometimes tables of direct products are given in which the antisymmetrized representation is distinguished by being bracketed. In the current example $T_2 \times T_2 = A_1 + E + [T_1] + T_2$

Table 4.9 Direct products $T_2 \times T_2$ in the point group T_d

R	E	$8C_3$	$3C_2$	$6S_4$	$6\sigma_d$
T_2	3	0	-1	-1	1
$T_2 \times T_2$	9	0	1	1	1
$\chi(R^2)$	3	0	3	-1	3
$\chi^{(+)}$	6	0	2	0	2
$\chi^{(-)}$	3	0	-1	1	-1

A special case of the direct product of a representation with itself arises when the functions f and g are the same. The direct product then only consists of the symmetrized part, which has dimension $\frac{1}{2}n(n+1)$, because the product functions $f_i f_j$ and $f_j f_i$ are identical. The antisymmetrised direct product disappears in this case because the antisymmetrized basis functions $f_i f_j - f_j f_i$ are all zero. This fact has important consequences for the degeneracies of electronic states and excited vibrational states, as we shall see in the next chapters and in II.

References

1. Bunker, P. R. (1979). *Molecular symmetry and spectroscopy*. Academic Press, San Diego.
2. Heine, V. (1964). *Group theory in quantum mechanics*. Pergamon Press, Oxford.
3. Bishop, D. M. (1973). *Group theory and chemistry*. Clarendon Press, Oxford.
4. Cotton, F. A. (1963) *Chemical applications of group theory*. Wiley, New York.
5. Schonland, D. S. (1965). *Molecular symmetry*. van Nostrand Reinhold, New York.
6. Atkins, P. W. (1983). *Molecular quantum mechanics*. 2nd edn. Oxford University Press, Oxford.
7. Weyl, H. (1950). *The theory of groups and quantum mechanics*. Dover, New York.
8. Wigner, E. P. (1959). *Group theory and its applications to the quantum mechanics of atomic spectra*. Academic Press, New York.
9. Atkins, P. W., Child, M. S., Phillips, C. S. G. (1970). *Tables for group theory*. Oxford University Press, Oxford.

5 Normal modes of vibration

5.1 Symmetries and normal coordinates

Normal coordinates for the vibration of a polyatomic molecule generate an irrep of the molecular point group, and thus vibrations can be given symmetry labels.

In Section 2.9 we showed how to find the vibrational frequencies of a molecule. To do this we need to know the second derivatives of the potential energy with respect to the nuclear coordinates, since these are the force constants for the vibrational motions. At the end of the calculation we will have obtained a set of vibrational frequencies with a normal coordinate corresponding to each. We may find that some of the vibrations have the same frequency. The appearance of such a degeneracy is not usually an accident. Normal coordinates with the same frequency are generally symmetry-related—in the terms of Chapter 4 they form a basis for a degenerate irrep of the point group. The reason underlying this observation is that the potential energy of the molecule is invariant to all symmetry operations of the molecular point group. If a molecule is distorted in some way then a symmetry transformation from the point group of the equilibrium geometry must lead to a similar distortion with the same energy. If the force constant for a particular motion is k then the force constant for the transformed motion must also be k. For example, the force constant for distorting the methane molecule by stretching one of the C–H bonds must be the same as the force constant for stretching any of the other bonds.

The mass-weighted normal coordinates introduced in Section 2.9 are derived so that around the equilibrium configuration the potential energy takes the form

$$V = \sum_{\alpha,i} \omega_\alpha^2 Q_{\alpha i}^2 \tag{5.1}$$

The sum over α covers all vibrational frequencies and the sum over i covers all the normal coordinates $Q_{\alpha i}$ which are associated with the particular frequency ω_α. Now consider a molecule distorted from its equilibrium configuration along one of these normal coordinates. If the distortion is small the potential energy is proportional to the square of the displacement. The effect of a symmetry operation is to transform the distortion to a different (symmetry-related) direction, but the symmetry operation cannot alter the energy of the molecule. The potential energy of the transformed molecule therefore also depends quadratically on the distortion, with the same constant of proportionality ω_α^2. Any symmetry transformation of the normal coordinate therefore transforms it to a linear combination of normal coordinates with the same frequency. In consequence a set of normal coordinates with the same frequency acts as a basis for a representation of the point group.

If a vibration is nondegenerate no other normal vibration has the same frequency and so the symmetry operation must transform the normal coordinate into itself. The normal coordinate must therefore be a basis for a one-dimensional irrep. For a vibration with degeneracy two, at least one symmetry operation mixes the normal coordinates, otherwise there is no symmetry reason for them to have the same frequency. Thus we conclude that the normal coordinates are a basis for a two-dimensional irrep. Similarly, that a set of n independent normal coordinates with a given frequency is a basis for an n-dimensional irrep.

5.2 Finding the symmetries of the vibrations

The basis functions

The first step in the method for finding the symmetries of the normal modes of vibration is to set up a basis. This will be a coordinate system which can be used to describe all the possible ways the molecular geometry can be distorted. Many coordinate systems can be used, but the simplest applies a set of Cartesian axes to describe the displacement of each atom in the molecule from its equilibrium position. Three coordinates (x, y, z) are required to measure the displacement of each atom from its equilibrium position, thus the basis contains a total of $3N$ coordinates. The $3N$ dimensions cover the $3N - 6$ vibrational coordinates describing the distortions of the molecule and also the coordinates of the centre of mass, and the orientation of the molecule in space. These three-dimensional sub-spaces describe the translational and rotational motions of the molecule, and can be separated out following reduction of the representation, leaving the $3N - 6$ normal vibrational coordinates.

The method will be illustrated using methane as an example. The coordinate system employed is depicted in Fig. 5.1.

The basis used is a set of $3N$ coordinates which describe every possible way of moving the N atoms in the molecule. The representation obtained contains all the possible motions of the molecule: vibration, rotation and translation.

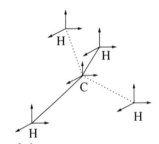

Fig. 5.1 Displacement coordinates for methane

The representation

The next stage is to use the basis to generate characters for a representation of the group. As in Chapter 4 the characters are worked out by considering the effect on the basis coordinates of one symmetry operation in each class. If an atom is moved by a symmetry operation, its coordinate system is transformed to a system based on a different location, and contributes nothing to the character. The only atoms that are not moved lie on the symmetry element and the symmetry operation has the same effect on all such coordinate systems.

For example, reflection in the σ_d plane passing through H atoms 1 and 2 in Fig. 5.2 will leave both these atoms and the central C atom unmoved, whereas atoms 3 and 4 will be interchanged. The coordinate systems for atoms 3 and 4 therefore contribute nothing to the character. The other three atoms each have coordinate systems whose origins lie in the σ_d plane. Reflection has the same effect on all three sets of axes. If they are oriented as in Fig. 5.2, then on reflection the unit vectors along the x and y directions are interchanged, giving zero contribution to the character, whereas the vector along the z axis is invariant. Thus, each axis system gives a total contribution to the character of 1, and the

A symmetry operation has the same effect on all axis systems whose origin lies on the symmetry element.

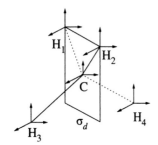

Fig. 5.2 σ_d operation for methane

character for the σ_d class is 3 (1 for each atom).

We see that the representation can be generated in two ways:

1. Consider the whole basis and determine the effects of the symmetry operations (representation $\Gamma^{(d)}$, d for displacement).

2. Consider two smaller representations: the first generated by counting the number of atoms that do not move for each symmetry operation, ($\Gamma^{(a)}$, a for atoms), and the other generated by an axis system placed at the centre of mass of the molecule, which is static in all symmetry operations ($\Gamma^{(t)}$, t for translation). The required representation is obtained by multiplying the character of each symmetry operation in $\Gamma^{(t)}$ by the number of atoms not moved by the operation, the *direct product* $\Gamma^{(a)} \times \Gamma^{(t)}$ (see Section 4.9).

Both methods are perfectly satisfactory. In practice the second is easier, since to find $\Gamma^{(a)}$ we simply count the atoms that do not move, and for $\Gamma^{(t)}$ there is only one coordinate system to worry about. $\Gamma^{(t)}$ is usually identified in character tables, in rows labelled x, y, z as appropriate (sometimes T_x, T_y, T_z). To obtain the characters of $\Gamma^{(d)}$ we multiply the characters of the two smaller representations. The characters for the methane molecule are given in Table 5.1.

Table 5.1 Representation generated by atomic displacements in methane

R	E	$8C_3$	$3C_2$	$6S_4$	$6\sigma_d$
$\Gamma^{(d)}$	15	0	-1	-1	3
$\Gamma^{(a)}$	5	2	1	1	3
$\Gamma^{(t)}$	3	0	-1	-1	1

$\Gamma^{(d)}$ can be decomposed using the reduction formula, eqn 4.28, giving

$$\Gamma^{(d)} = A_1 + E + T_1 + 3T_2 \qquad (5.2)$$

There are four triply degenerate irreps, one doubly degenerate irrep and one nondegenerate irrep. However, this 15-dimensional representation includes translations and rotations as well as internal motions. Our next task is to identify the symmetries of the translations and rotations and to remove them, leaving the symmetries of the 9 vibrational coordinates.

Translations and rotations

The symmetries of the translations and rotations are identified in standard character tables.

In practice it is very easy to identify translations and rotations, since their symmetries are given in the character table. Since the potential energy of a molecule is not altered by moving its centre of mass, translation is represented by a set of normal coordinates with a frequency of zero. The normal translational coordinates describe the motion of the centre of mass, and can be represented by displacements in a coordinate system at the centre of mass of the molecule. We have just considered the representation generated by this basis, it is $\Gamma^{(t)}$, and in the case of the methane molecule it accounts for one of the T_2 irreps we have found.

Changing the orientation of a molecule does not alter its potential energy either, and so the rotational normal coordinates are also associated with a force constant of zero. Rotations are represented by a set of orthogonal vectors based at the centre of mass. However, these (axial) vectors are different from normal (polar) vectors, because they represent angular momentum. Their direction is determined from the direction of rotation by the right-hand screw rule. The difference in transformation properties can be illustrated by considering the reflection in the xy plane of a vector directed along the x axis, as illustrated in Fig. 5.3. The x axis points out of the plane of the paper and the arrows indicate the direction of rotation. Clearly the rotation changes from clockwise to anticlockwise in the reflection, and so the angular momentum reverses its direction, whereas a polar vector along the x direction would be invariant to the reflection. In the point group T_d three orthogonal axial vectors based on the centre of mass have T_1 symmetry. However, it is not usually necessary to work this out—the representation generated by the three rotational degrees of freedom, $\Gamma^{(r)}$, is identified in published character tables, where it is is labelled R_x, R_y and R_z. For methane, a glance at the T_d character table (Table 4.6, page 68) shows that the rotations have T_1 symmetry. The triple degeneracy is an indication that three principal components of the moment of inertia are equal so that methane is a spherical top (see Section 3.5).

The symmetries of the vibrations are found by subtracting the translations and the rotations from $\Gamma^{(d)}$:

$$\Gamma^{(v)} = \Gamma^{(d)} - \Gamma^{(t)} - \Gamma^{(r)} \tag{5.3}$$

In methane the vibrational symmetries are therefore $A_1 + E + 2T_2$. Even though there are nine normal vibrations there are only four distinct vibrational frequencies—one is nondegenerate (A_1), one is doubly degenerate (E), and two are triply degenerate ($2T_2$). The forms of the A_1 and E normal coordinates can be found by using projection operators, but to separate the two T_2 bases requires more detailed analysis, because they have the same symmetry. Note, however, that the number of distinct frequencies, their symmetries and degeneracies have been found from group theory without any knowledge of the force constant matrix whatsoever. Group theory is a vital tool for the analysis of vibrations.

Stretching vibrations

The preceding method gives the symmetries of all the normal modes of vibration of a molecule. It is often useful to concentrate on the part of an infra-red spectrum where the vibrations can be approximately characterized as bond stretches. For example, inorganic carbonyl compounds have characteristic infra-red absorptions around 2000 cm^{-1} which give important structural information. In fact, it is generally easier to find the symmetries of the stretching vibrations than it is to find all the normal coordinate symmetries. If we use a basis of displacement coordinates representing distortion of the molecule along the equilibrium bond

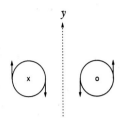

Fig. 5.3 Reflection of an axial vector. The arrows indicate the direction of rotation.

Problem 5.2.1. *Find the symmetries and degeneracies of the normal vibrational coordinates of the following species: NH$_3$, cis- and trans-CHF:CHF, AuCl$_4^-$.*

To find the symmetries of the bond stretching vibrations the basis coordinates are unit vectors directed along the equilibrium bond directions.

Problem 5.2.2. *Generate the representation for the stretching vibrations of methane and show that it is equal to $A_1 + T_2$*

Table 5.2 Vibrations of methane

Symmetry	ω_e / cm^{-1}
A_1	3143
E	1534
T_2	3019
T_2	1306

Problem 5.2.3. *The CO stretching vibrations of metal carbonyls can be used to distinguish between isomers. Find the symmetries of the C=O stretching modes of both mer- and fac-isomers of the type $ML_3(CO)_3$.*

Because the normal coordinates of vibration generate irreps of the molecular point group, vibrational wavefunctions can also be given symmetry labels.

The ground state vibrational wavefunction is totally symmetric.

directions we will find the appropriate representation immediately.

In the methane molecule the stretching vibrations are motions of the H atoms along their equilibrium directions. We therefore choose displacement coordinates directed along the bonds. To generate the representation we simply count how many of the bonds are invariant for each symmetry operation. Thus, for example the reflection in the σ_d plane containing H atoms 1 and 2, depicted in Fig. 5.2, leaves C–H bonds 1 and 2 invariant, and interchanges C–H bonds 3 and 4. The character in the representation is therefore 2.

In a problem like this we do not need to subtract translations and rotations, since the basis only includes distortions along the bond directions. Neither translation nor rotation involves such distortions. We therefore conclude that the A_1 vibration and one of the T_2 vibrations of methane are C–H stretches, which we expect to find around 3000 cm^{-1}. The vibrational frequencies of methane and their symmetries are given in Table 5.2.

5.3   Symmetries of vibrational wavefunctions

As shown in Section 2.9 the vibrational wavefunctions of polyatomic molecules can be expressed as products of simple harmonic oscillator wavefunctions, each of which is a function of one normal coordinate. Each simple harmonic oscillator wavefunction, ϕ_i, has a characteristic frequency ω_i and a vibrational quantum number v_i. The energy of the molecule relative to the zero-point energy level is given by the sum of all the separate harmonic oscillator energies, i.e.

$$E = \sum_i v_i \hbar \omega_i \qquad (5.4)$$

The quantum number for each normal mode is interpreted as the number of quanta in the vibration.

The ground state

For the sake of understanding vibrational spectra it is important to clarify the relationships between the symmetries of the normal coordinates and the symmetries of the possible vibrational wavefunctions. We start with the vibrational ground state. Reference to Sections 2.9 and 3.3 shows that the ground state wavefunction is of the form (cf eqn 3.14).

$$\psi = N_0 \exp(-\sum_{\alpha,i} \omega_\alpha Q_{\alpha i}^2 / 2\hbar) \qquad (5.5)$$

The sum $\sum_i Q_{\alpha i}^2$ over all normal coordinates with frequency ω_α is invariant to every symmetry operation of the point group because the set of normal coordinates with a given frequency forms a basis for a representation of the group. If these normal coordinates are arranged in a vector, $\mathbf{Q}_\alpha$, then the sum can be written as the scalar product

$$\sum_i Q_{\alpha i}^2 = \mathbf{Q}_\alpha^T \mathbf{Q}_\alpha \qquad (5.6)$$

which is invariant to all symmetry operations, as proved in Section 4.1. In consequence, the exponent in eqn 5.5 is invariant to all symmetry operations and therefore the ground vibrational wavefunction is totally symmetric.

Fundamental states

In the *fundamental* states, only one normal mode is excited with a single quantum. If mode j is excited, the wavefunction has the form (see Table 3.1)

$$\psi = N_1 Q_j \exp(-\sum_{\alpha,i} \omega_\alpha Q_{\alpha i}^2 / 2\hbar) \qquad (5.7)$$

As we have just shown, the exponential term is totally symmetric, and so the wavefunction must have the same symmetry as the normal coordinate Q_j.

If there is a single quantum in a vibration, the vibrational wavefunction has the same symmetry as the normal coordinate.

Combination states

There are also higher excited vibrational states with several quanta of excitation, for example the *combination* states in which two vibrations are excited with one quantum in each. The symmetry of the resulting state is obtained from the direct product of the two constituent normal mode symmetries. If one of the normal modes is degenerate the resulting direct product representation may also be degenerate. For example the methane molecule has two vibrations of symmetry T_2 but different frequencies. In the possible combination states, where there is one quantum in each, the resulting wavefunctions generate the representation $T_2 \times T_2 = A_1 + E + T_1 + T_2$. There is no symmetry reason for all these nine states to be mutually degenerate, and once anharmonic corrections have been applied we find four energy levels with degeneracies 1, 2, 3 and 3, as indicated by group theory.

Overtone states

Another important type of higher state is the first *overtone* state, in which a given vibrational frequency is excited with two quanta. In the case of a nondegenerate vibration the direct product must give a totally symmetric state. Overtones of degenerate vibrations are more complicated. Because we are now considering the representation generated by the direct product of a basis with itself, the antisymmetrized direct product is identically zero, leaving us with only the symmetrized direct product, as discussed in Section 4.9. This result can be understood qualitatively in the following way. The possibilities for the first overtone of an n-fold degenerate vibration are to put two vibrational quanta into the same normal coordinate, or to put one in each of two distinct degenerate coordinates. The total number of ways of doing this is n for the former possibility and $\frac{1}{2}n(n-1)$ for the latter, giving a total of $\frac{1}{2}n(n+1)$. In the case of the overtone of a T_2 vibration in methane, the symmetrized direct products were worked out in Section 4.9 and are $A_1 + E + T_2$.

The detailed analysis of the symmetry of higher overtone states for degenerate vibrations is complicated. The total number of possibilities for a vibration of degeneracy n containing m quanta is [5] $(n + m - 1)!/ (n - 1)!m!$. This formula is important in the theory of unimolecular reactions, since it represents the approximate degeneracy of a highly excited vibrational energy level.[6]

5.4   Vibrational spectra

Infra-red spectrum

Vibrations are infra-red active if they have the same symmetry as some component of the dipole moment operator, $\Gamma^{(t)}$

If a molecule absorbs light of an appropriate frequency (usually in the infra-red region) it is excited from one vibrational state to another. The strongest bands in the spectrum generally correspond to excitations from the ground state to a fundamental state (i.e. to excitation of one normal mode only). However, not all the possible fundamental states can be produced in this way, and so not all the fundamental vibrational frequencies appear in the infra-red spectrum.

Light consists of an oscillating electric field, which interacts with an oscillating dipole moment in the molecule. Thus, on a simple level, only vibrations in which the molecular dipole moment changes with the vibrational motion can be seen in the spectrum. This idea can be made more precise by using the methods of Section 1.7. The excited wavefunction $|i>$ and the ground state wavefunction $|0>$ are mixed by the interaction of the light with the electric dipole moment of the molecule. The dipole moment operator is a vector with components $\hat{\mu}_x$, $\hat{\mu}_y$ and $\hat{\mu}_z$, and therefore has the same symmetry as a vector based on the centre of mass, $\Gamma^{(t)}$, as discussed above. The intensity of the transition is proportional to the quantity

$$\sum_{k=1}^{3} |<i|\hat{\mu}_k|0>|^2 \qquad (5.8)$$

If all three integrals $<i|\hat{\mu}_k|0>$ are zero then the vibration does not alter the dipole moment of the molecule in any direction. The transition is then symmetry-forbidden, and we do not expect to see it in the infra-red spectrum. If at least one of these integrals is nonzero then the vibration is said to be infra-red active. But we can tell whether there is a symmetry reason for these integrals to vanish by considering whether the direct product representation $\Gamma^{(i)} \times \Gamma^{(t)} \times \Gamma^{(0)}$ contains the totally symmetric irrep. Since the ground state wavefunction is totally symmetric we have

$$\Gamma^{(i)} \times \Gamma^{(t)} \times \Gamma^{(0)} = \Gamma^{(i)} \times \Gamma^{(t)} \qquad (5.9)$$

Following the discussion of vanishing integrals in Section 4.8, a necessary and sufficient condition for the integrand to be totally symmetric is for the representation $\Gamma^{(t)}$ to contain the irrep $\Gamma^{(i)}$, in other words, for the vibrational symmetry to be the same as some component of the dipole moment operator. For the methane molecule $\Gamma^{(t)} = T_2$, and hence only the two triply degenerate vibrations of symmetry T_2 are infra-red active. The two remaining vibrations, of symmetry A_1 and E, are not observed in the infra-red spectrum.

In a molecule with a unique figure axis it is also possible to use group theory to distinguish *parallel* and *perpendicular* vibrations. A parallel vibration causes the dipole moment to oscillate parallel to the figure axis. It therefore has the same symmetry as the z component of the dipole moment operator. A perpendicular vibration has the same symmetry as the x and y components of the dipole moment operator. It is easy to distinguish parallel and perpendicular vibrations in the infra-red spectrum because they have different rotational fine structures,[3] as can be seen for bromomethane in Fig. 5.4.

Problem 5.4.1. *Find the symmetries of the normal modes of vibration for the bromomethane molecule, identify those which are infra-red active and classify them as parallel or perpendicular.*

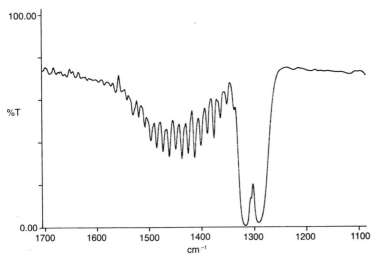

Fig. 5.4 Parallel and perpendicular vibrations in the infra-red spectrum of bromomethane. The band with a smooth double hump appearance centred around 1300 cm^{-1} is characteristic of a parallel vibration (A_1 symmetry). The band which appears as a series of spikes around 1450 cm^{-1} is characteristic of a perpendicular vibration (E symmetry). The differences come from the different rotational selection rules associated with each type of transition, see reference 3.

Raman spectrum

In the vibrational Raman spectrum a photon is scattered off the molecule with a change in frequency and the molecule is left in a new vibrational state. This is a two-photon process, whose intensities depend on the polarizability operator $\hat{\alpha}$. The intensity of a transition is proportional to the quantity

$$\sum_{j,k=1}^{3} |<i|\hat{\alpha}_{jk}|0>|^2 \qquad (5.10)$$

Vibrations are Raman active if they have the same symmetry as some component of the polarizability operator, the symmetrized direct product $\Gamma^{(t)} \times \Gamma^{(t)}$

Problem 5.4.2. *The pattern of vibrations found in the infra-red and the Raman spectrum is an important piece of structural information. How could you expect to distinguish mer- and fac-$ML_3(CO)_3$ on the basis of the CO stretching frequencies found in each spectrum? (See Problem 5.2.3).*

Problem 5.4.3. *Find the symmetries of the vibrations of cis- and trans-CHF:CHF and decide which of them should be infra-red active, and which Raman active.*

The symmetry of the polarizability operator is the same as the successive application of the dipole moment operator twice, emphasizing that the Raman effect is a two-photon process. The process can be thought of as absorption of a photon to a *virtual state* followed by rapid emission of another photon. The polarizability operator generates the symmetrized direct product representation $\Gamma^{(t)} \times \Gamma^{(t)}$, which is the same as the representation generated by the quadratic combinations of the Cartesian coordinates, x^2, y^2, z^2, xy, xz and yz. The relevant representations are identified in published character tables. The intensity of the Raman line is zero by symmetry unless the matrix element of some component of the polarizability is totally symmetric, i.e. unless the excited vibrational state has the same symmetry as a component of the polarizability. In the point group T_d we have already worked out the symmetrized direct product $T_2 \times T_2 = A_1 + E + T_2$. We therefore conclude that all the vibrations of methane are Raman active.

The result of Problem 5.4.3 illustrates the so-called *mutual exclusion rule*. It will be noticed that no vibration of *trans*-CHF:CHF is both infra-red active and Raman active. This observation illlustrates a general rule for molecules with a centre of symmetry, and arises because the dipole moment operator has u symmetry with respect to inversion, whereas the polarizability operator has g symmetry. Thus only vibrations with u symmetry are infra-red active, and only vibrations with g symmetry are Raman active. It is not possible to be both g and u at the same time since they designate opposite transformation properties with respect to inversion—thus no vibration can be found in both spectra if the molecule has a centre of symmetry.

References

1. Wilson, E. B., Decius J. C. and Cross, P. C. (1980). *Molecular vibrations*. Dover, New York.
2. Woodward, L. A. (1972). *Introduction to the theory of molecular vibrations and vibrational spectroscopy*. Oxford University Press, Oxford.
3. Herzberg, G. (1945). *Infrared and Raman spectra of polyatomic molecules*. van Nostrand Reinhold, New York.
4. Califano, S. (1976). *Vibrational states*. Wiley, New York.
5. Feller, W. (1968). *An introduction to probability theory and its applications*, 3rd ed. Wiley, New York.
6. Pilling, M. J. and Seakins, P. W. (1996) *Reaction kinetics*. Oxford University Press, Oxford.

6 Molecular orbitals

We now turn to consideration of the electronic wavefunction, which determines the distribution of electrons in the molecule, and therefore the bonding. Chemists describe bonding in terms of atomic and molecular orbitals, and in this chapter we show how group theory can be used to classify the symmetries of orbitals and to find suitable combinations of atomic orbitals for the construction of a molecular orbital. We start by considering the case of atoms before proceeding to the more complex problem of molecules.

6.1 Atoms

In the orbital approximation, introduced in Section 2.7, the electronic wavefunction of an atom is constructed as a product of one-electron wavefunctions, or orbitals. Each orbital obeys a Schrödinger equation of the form

$$-\frac{\hbar^2}{2m}\nabla^2\psi + V(\mathbf{r})\psi = \epsilon\psi \qquad (6.1)$$

where $V(r)$ is the potential energy, including the average potential energy of interaction with all other electrons in the atom.

For atoms it is usual to average the potential energy over all orientations of the nucleus–electron vector so that $V(r)$ becomes spherically symmetrical. This additional approximation is known as the *central field approximation*. Equations of this type have many possible solutions, each corresponding to a different atomic orbital with a different, self-consistent, spherically averaged $V(r)$, and each with a different orbital energy. However, the angular parts of the solutions are always spherical harmonics because of the spherical symmetry of the potential energy (see Sections 2.3 and 3.4). Each solution can therefore be labelled by an angular momentum quantum number l, and has degeneracy $2l+1$, the number of possible values for the quantum number m_l. The radial functions, however, are not the same as for the hydrogen atom, because $V(r)$ is not a simple Coulomb potential.

These familiar ideas can also be looked at from the point of view of group theory. In the central field approximation each one-electron Schrödinger equation is invariant to all symmetry operations of a sphere, which form the point group R_3. The set of orbitals with a given energy therefore forms a symmetry-related set which is a basis for an irrep of R_3. Some properties of the irreps of R_3 are given in Table 6.1. The symmetry labels for the irreps are S, P, D, F, etc. with degeneracies 1, 3, 5, 7, respectively. The characters of these irreps define the symmetry properties of the spherical harmonics, and thus each energy level (often

In the central field approximation each electron in an atom feels a spherically symmetrical potential energy. The angular dependence of the orbitals is therefore the same as in the H atom.

Table 6.1 Degeneracies of the spherical group R_3

Irrep	Degeneracy	l
S	1	0
P	3	1
D	5	2
F	7	3
G	9	4
...	$2l+1$	l

called a *sub-shell*) can be given a symmetry label indicating these symmetry properties, such as s, p, d, f etc. A detailed description of the electronic wavefunctions of atoms can be found in reference 1.

6.2   The LCAO method

In a molecule the potential energy is assumed to have the full symmetry of the molecular point group. The molecular orbitals can then be labelled with irreps of the group, which determine their symmetry properties.

For molecules the situation is more complicated because there are several nuclei acting as centres of attraction. In the Born–Oppenheimer approximation, where the nuclei form a rigid framework, the one-electron Schrödinger equation, eqn 6.1 is not spherically symmetrical. The electronic kinetic energy operator always has spherical symmetry. However, the potential energy of the electron reflects the symmetry of the nuclear arrangement, and is assumed to be invariant to symmetry operations of the molecular point group. This assumption takes the place of the central field approximation for atoms. In consequence a set of degenerate orbitals forms a basis for an irrep of the molecular point group.

It is normal to distinguish between *valence electrons* and *core electrons*. Chemical bonding arises from the rearrangement of the valence electrons, to occupy molecular orbitals extending over several nuclei. To a first approximation the core electrons do not take part in the bonding and can still be associated with a single nucleus. It is natural to try to construct the molecular orbitals by superposing appropriate combinations of the valence atomic orbitals (the LCAO method). Thus, if we denote the molecular orbital by ψ and the ith valence atomic orbital by ϕ_i then we construct a molecular orbital by linear combination of the atomic orbitals:

$$\psi = \sum_i a_i \phi_i \qquad (6.2)$$

where a_i are the coefficients in the linear combination. There are infinitely many linear combinations of this form for even the smallest molecule. We clearly need some way of finding the correct combinations. As with the case of the normal vibrational coordinates this procedure can be made much simpler by the use of group theory.

Since each set of degenerate molecular orbitals must generate an irrep of the molecular point group, the problem will be greatly simplified if we can find those combinations of the valence atomic orbitals that generate each possible irrep. As in the case of vibrations, group theory does not give energies, but it does give the symmetries and the degeneracies of the orbitals. The method is now familiar.

1. The basis functions are chosen to be the valence orbitals of the atoms that make up the molecule.

2. The basis is used to generate a representation of the point group.

3. The representation is reduced.

4. Projection operators are applied to determine the *symmetry-adapted linear combinations*.

As an example we consider the methane molecule. The valence orbitals are the four $1s$ orbitals on the H atoms, the three $2p$ orbitals and the $2s$ orbital on the C atom. Each of these three sets of orbitals would, on its own, form a basis for a representation of T_d. For example, the carbon $2p$ orbitals are transformed into one another by the symmetry operations, but there is no symmetry operation that can transform a C $2p$ orbital into a H $1s$ orbital. We have already considered the four H $1s$ orbitals in Section 4.4 where we showed that they generate the representation $A_1 + T_2$. The C $2s$ orbital is totally symmetric, since it has spherical symmetry and is located at the centre of mass. The $2p$ orbitals have the same symmetry as an (x, y, z) axis system based at the centre of mass, and therefore generate $\Gamma^{(t)} = T_2$ (see Section 5.2). The total representation is therefore $2A_1 + 2T_2$. We conclude that the molecule will have two nondegenerate a_1 molecular orbitals and two triply degenerate t_2 sets of molecular orbitals in the valence shell. The two a_1 orbitals will be superpositions of the C $2s$ orbital and the A_1 combination of the four H $1s$ orbitals (which is $H_1 + H_2 + H_3 + H_4$). We expect that one of these combinations will be bonding (i.e. the orbitals will interfere constructively and be lowered in energy); the other combination will be raised in energy and be an antibonding orbital (destructive interference). Similar considerations will apply for the two t_2 sets of orbitals. The exact coefficients are not completely determined by symmetry, except in rare cases, and will be considered in more detail in II. A schematic orbital correlation diagram is shown in Fig. 6.1. Only combinations of AOs with the same symmetry can be combined with one another to form an MO because each MO must be a basis for an irrep. Thus an MO of A_1 symmetry can only be built from the C $2s$ orbital, and the A_1 combination of the H orbitals.

The eight valence electrons are distributed among the molecular orbitals in accordance with the Aufbau principle, filling the lowest four MOs, which are the bonding a_1 orbital and the triply degenerate bonding t_2 orbital. The antibonding orbitals are all empty, giving a total of eight bonding electrons (or equivalently four CH bonds), in agreement with the usual picture of the methane molecule. Notice, however, that the simple molecular orbital theory predicts that the bonding electrons are divided between orbitals with two different energies, two in the a_1 orbital and six in the t_2 orbitals. The qualitative value of the MO picture is reinforced by an analysis of the photoelectron spectrum, Fig. 6.2, in which the lowest ionization energy is interpreted as ionization from a triply degenerate orbital. (Strictly speaking the orbital symmetry of the resulting ion is T_2, see II.) Notice also that this simple molecular orbital picture seems to contradict the idea of sp^3 hybridization. No mixing of the s and p orbitals is possible in methane because they have different symmetries.

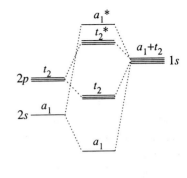

Fig. 6.1 Molecular orbital correlation diagram for methane. The dashed lines show the correlations between symmetry adapted combinations of atomic orbitals and molecular orbitals of the same symmetry.

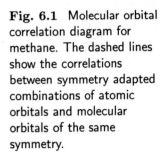

Fig. 6.2 Photoelectron spectrum of methane.

Problem 6.2.1. *Find the symmetries of the valence molecular orbitals in the LCAO approximation for the following molecules, and propose an orbital correlation diagram for each: NH_3, O_2, H_2O.*

6.3 π bonding

The ideas of molecular orbitals and delocalization of electrons across

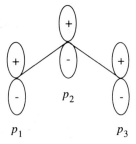

Fig. 6.3 Basis functions for π bonding in the allyl radical.

Table 6.2 Transformations of the out of plane p orbitals in the allyl radical

E	C_2	σ_v	σ_v'
p_1	$-p_3$	p_3	$-p_1$
p_2	$-p_2$	p_2	$-p_2$
p_3	$-p_1$	p_1	$-p_3$

Table 6.3 Representations of the π orbitals in the allyl radical

C_{2v}	E	C_2	σ_v	σ_v'
$\Gamma^{(13)}$	2	0	0	-2
$\Gamma^{(2)}$	1	-1	1	-1
A_1	1	1	1	1
A_2	1	1	-1	-1
B_1	1	-1	1	-1
B_2	1	-1	-1	1

Problem 6.3.1. *Verify the characters in Table 6.3 and the reduction of the two representations $\Gamma^{(2)}$ and $\Gamma^{(13)}$. Apply projection operators to show (using Table 6.2) that the A_2 combination of p_1 and p_3 is $p_1 - p_3$ and that the B_1 combination is $p_1 + p_3$.*

several atoms are probably most familiar in the context of π bonding and aromaticity. We shall therefore consider the problem of finding the symmetries and degeneracies of the π orbitals in a delocalized system.

One of the simplest delocalized π systems is found in the allyl radical, $CH_2=CH-CH_2^{\bullet}$, which belongs to the C_{2v} point group. There are two equivalent structures which can be considered to be in resonance, one of which has the odd electron on the right hand C atom and the other has it on the left. The σ bonding orbitals of the carbon framework are constructed from the carbon $2s$ orbitals and the $2p$ orbitals that lie in the plane of the molecule. The π bonding orbitals are constructed from the $2p$ orbitals that are orthogonal to the plane of the molecule. These orbitals are also orthogonal to the σ bonding orbitals because they have opposite symmetries with respect to reflection in the plane of the molecule. They can therefore be considered independently of the σ bonding framework.

To illustrate the use of group theory here we consider the effect of each symmetry operation of the C_{2v} point group on the three out-of-plane $2p$ orbitals, shown in Fig. 6.3. The ways in which these three orbitals are transformed are tabulated in Table 6.2. It will be seen that orbitals 1 and 3 are interconverted by some symmetry operations but orbital 2 is transformed into itself by each operation, because the central atom is not related by symmetry to the two outer atoms. We therefore consider the basis made up of orbitals 1 and 3 separately from the basis made up of orbital 2. The representations generated by each of these bases are given in Table 6.3 together with the C_{2v} character table. It can be seen, either by inspection or by use of the reduction formula, that $\Gamma^{(2)} = B_1$ and that $\Gamma^{(13)} = A_2 + B_1$.

We conclude that there are three π orbitals in the valence shell of the allyl radical. One of them has A_2 symmetry and is equal to $p_1 - p_3$. The other two have B_1 symmetry and are combinations of $p_1 + p_3$ with p_2. One of these will be bonding and the other will be antibonding. The a_2 orbital is essentially nonbonding as there is a node on atom 2. An energy level diagram is given in Fig. 6.4. The allyl radical has three π electrons. The first two occupy the b_1 bonding orbital and the odd electron goes into the nonbonding a_2 orbital. It therefore has significant density only on the terminal C atoms, a fact borne out by the tendency of the allyl radical to react at the terminal C and by the esr spectrum, which shows hyperfine coupling between the odd electron and the two terminal CH_2 groups, but none with the central CH group.

Problem 6.3.2. *Perform a similar analysis for the cis-butadiene molecule to show that the π molecular orbitals have symmetries $2a_2 + 2b_1$. Identify the a_2 and b_1 combinations of the orbitals on atoms 1 and 4 and on atoms 2 and 3. By mixing combinations of the same symmetry in bonding and antibonding ways suggest the order of energies for the π molecular orbitals, sketch them and identify the HOMO and the LUMO.*

Problem 6.3.3. *In benzene all the carbon atoms are symmetry-related.*

Show the π orbitals of benzene, and show that they have symmetries
$a_{2u} + b_{2g} + e_{2g} + e_{1u}$. *Apply projection operators to find the forms of these*
orbitals and hence predict the order of their energies and the ground state
configuration of the π electrons in benzene.

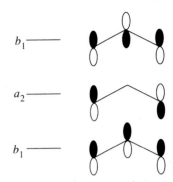

Fig. 6.4 MOs for the allyl radical.

6.4 Orbital symmetry and chemical reactivity

Molecular orbitals and their symmetries are not only useful in rationalizing the bonding of simple molecules, they can also be used to give insight
into the pathways of some chemical reactions. In analysing a reaction
pathway it is common to identify a transition state and to trace the
symmetries of the valence orbitals of the transition state from reactants
to products. If the valence orbitals of the reactants have the correct symmetries to interact and lead to the transition state, then the reaction is
said to be *symmetry-allowed*. If not, the reaction is *symmetry-forbidden*.

Chemical reactions involve transfers of electron density between orbitals, and in the *frontier orbital* method it is usual to concentrate simply
on the HOMO of one molecule and the LUMO of the other. The new
bonding comes from the interaction of these orbitals, which is favoured
if they have large overlap and similar energies. The method is not exact
for several reasons:

1. The transition state may not have any exact symmetry elements,
   even when the reactant molecules do. The first example we consider
   below is such a case.

2. The bonding in a transition state cannot be described accurately
   using only the HOMO and the LUMO.

3. Some reactions are not *adiabatic*. For a nonadiabatic reaction a
   transition occurs from one electronic state to another during the
   course of the reaction and the bonding does not therefore change
   smoothly from reactant to product.

In spite of these caveats orbital symmetry arguments have proved
particularly useful in understanding electrocyclic reactions and sigmatropic rearrangements, and we briefly consider one example of each. For
a more detailed exposition the reader is referred to reference 4.

The Diels–Alder reaction

The Diels–Alder reaction is an electrocyclic reaction between a conjugated diene, such as cyclopentadiene, and a *dienophile*, which can be
considered to be a two-electron π system, such as methyl acrylate. In
the frontier orbital picture the key interaction is between the HOMO of
the diene and the LUMO of the dienophile. The reaction is enhanced
by neighbouring groups which raise the HOMO energy of the diene
or reduce the LUMO energy of the dienophile, thus improving the energy matching. The most important features of the HOMO of the diene
and the LUMO of a typical single π bond are shown in Fig. 6.5. If the

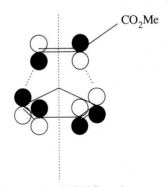

Fig. 6.5 HOMO and LUMO for a Diels–Alder reaction.

Problem 6.4.1. *Use the symmetry of the frontier orbitals to discuss the type of transition state for which a 2+2 cycloaddition, such as ethene + ethene is symmetry-allowed.*

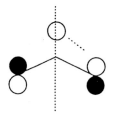

Fig. 6.6 Supra transition state for a 1–3 sigmatropic shift.

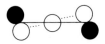

Fig. 6.7 Antara transition state for a 1–3 sigmatropic shift. There is a C_2 axis perpendicular to the plane of the paper.

Problem 6.4.2. *Show that the 1–5 sigmatropic shift in penta-1,3-diene is symmetry-allowed through a supra transition state.*

dienophile is considered to attack the diene symmetrically from above or below the plane of the conjugated pi system of the diene the transition state will have a plane of symmetry, as marked on the figure. The HOMO and LUMO have the same symmetry with respect to this plane, which is the only symmetry operation remaining in the transition state (in the case shown it is only approximately a symmetry operation). In consequence we conclude that the reaction is symmetry-allowed with this (supra–supra) transition state.

1–3 Sigmatropic shift of propene

Another type of reaction for which orbital symmetry arguments are helpful is the sigmatropic shift of an H atom from one C to another. A simple example of such a reaction is the shift of an H atom from one terminal C atom to the other in propene. Of course for propene the product of this reaction is identical to the reactant, but if the propene is substituted this is no longer the case. The simplest way to visualize the transition state for this reaction is to disconnect the H atom, leaving an allyl radical. The frontier orbital interactions are between the SOMO of the allyl radical (see Section 6.3) and the SOMO ($1s$ orbital) of the H atom. The orbitals and their interactions are shown for a supra transition state in Fig. 6.6. In order to transfer from one terminal C atom to the other the H $1s$ orbital must have positive interactions with both C atoms. This is not the case for the supra transition state. The SOMO of the allyl radical has the opposite symmetry to the H-$1s$ orbital with respect to the one remaining reflection symmetry. In order to have a symmetry-allowed reaction some way must be found for the H atom to interact with the lobes of the SOMO on opposite sides of the plane. This can be done in principle by twisting the two C atoms in the same direction and placing the H atom in the CCC plane on the C_2 axis, which is now the only remaining symmetry element. This *antara* transition state is depicted in Fig. 6.7. Although the reaction is symmetry-allowed through the antara transition state, so much steric strain must be introduced to twist the C atoms against the conjugated system that the reaction does not readily take place.

References

1. Woodgate, G. K. (1983). *Elementary atomic structure.* 2nd edn. Oxford University Press, Oxford.
2. Schonland, D. S. (1965). *Molecular symmetry.* van Nostrand Reinhold, New York.
3. Bishop, D. M. (1973). *Group theory and chemistry.* Clarendon Press, Oxford.
4. Fleming, I. (1982). *Frontier orbitals and chemical reactivity.* Wiley, New York.